# Thermochemical Reactions

*Numerical Solutions*

by D. James Benton

# Foreword

Thermochemistry is the science of analyzing molecular reactions to determine if they are spontaneous, energy absorbing or releasing, and to predict the product mole ratios and rates. Chemical reactions, like most other processes, tend to follow the path of free energy minimization or entropy maximization. This principle forms the mathematical basis for the analytical approach. This book is a how-to manual, filled with many examples and comes with all the code you need to accomplish this task.

## A Secret Hidden for 30 Years

I reveal in Chapter 8 the previously unpublished secret to solving the Gibbs' problem (i.e., free energy minimization constrained by elemental abundances) for nonideal reactants with vanishing fugacities. As the fugacity coefficient, $\varphi$, (or the compressibility, Z) of any reaction product decreases below unity, the Hessian (i.e., matrix containing the second partial derivatives of the total free energy, $\partial^2 G/\partial y_I/y_J$) becomes increasingly ill-conditioned and the steepest descent search toward the minimum becomes indeterminate, but...

## Free Software

*All of the examples contained in this book,*
*(as well as a lot of free programs) are available at...*

*https://www.dudleybenton.altervista.org/software/index.html*

## Programming

All of the examples presented in this book are implemented either in the C programming language or Excel® macros. The code, spreadsheets, data files, and other material are arranged in folders within a single ZIP archive that can be freely downloaded at the address above. An interactive Windows® chemical reaction solving tool (CREST described in Appendices A and B) is freely available at the site listed above.

## Units

Throughout this text we will primarily use SI units. The software described herein will use either SI or English units (often called US Customary). The practicing scientist/engineer must be comfortable with any and all systems of units. Debates as to which are superior and why are among the most pointless conversations ever pursued. My advice is simple: *Get over it!*

**Figure 1. Combustion in Beaker**

**Figure 2. Blue Flames**

# Table of Contents

**Figure 3. Chemicals**

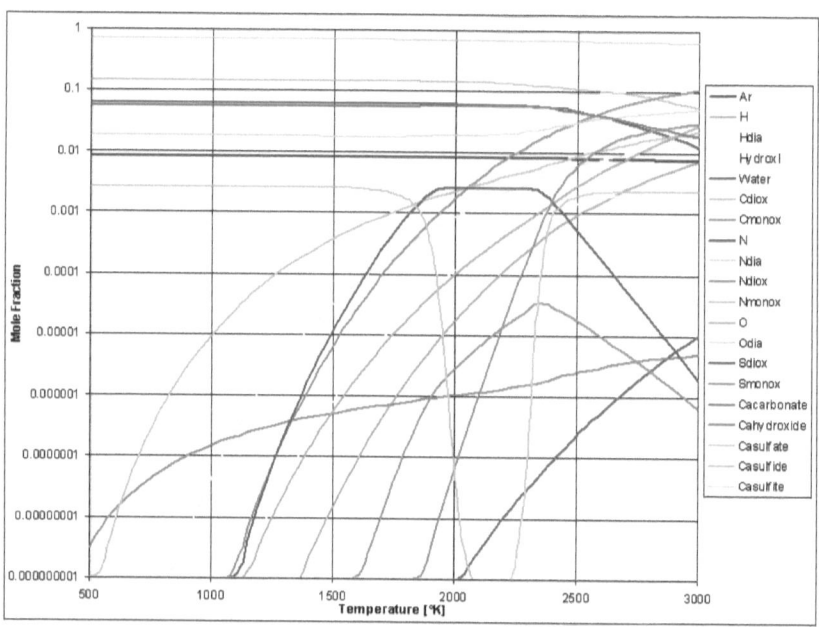

**Figure 4. Combustion of Coal with Air and Limestone to Capture Sulfur**

# Chapter 1. Introduction

Thermochemistry is the science of analyzing molecular reactions to determine if they are spontaneous, energy absorbing or releasing, and to predict the product mole ratios and rates. Chemical reactions, like most other processes, tend to follow the path of free energy minimization or entropy maximization. This principle forms the mathematical basis for the analytical approach.

Mixing of different chemical species, such as water and wine, changes both. Except for such things as water and oil—which don't mix—substances that do mix, will not separate spontaneously. Separation requires work—if it can be done at all. Chemical reactions produce different molecules or species. The species preceding the reaction are called *reactants* and those following or ensuing from the reaction are called *products*.

## Entropy and Free Energy

Entropy is sometimes described as the disorder of a substance or system. A more academic description for entropy would be the probability of the current state. In statistical thermodynamics, entropy is proportional to the ways in which a system can be arranged or the number available states. Conceptually, a perfect quartz crystal has only one state; whereas, quartz sand has innumerable states. This famous expression for entropy is engraved on Boltzmann's tombstone.[1]

$$s = k \log W \qquad (1.1)$$

Here, $s$ is entropy, $k$ is Boltzmann's constant[2], and $W$ is the number of microstates (small-scale structure) associated with the macro-state (large-scale structure). Boltzmann proposed this equation, but could never rigorously prove it, in part due to the fact that there is no practical way of counting (or measuring) microstates. Albert Einstein criticized Boltzmann for this on several occasions. No completely satisfactory proof has yet to arise and I will not attempt one here. From these descriptions and arguments we can at least say that all systems tend toward their most probable state, which is also that of maximum entropy.

After the discovery of energy as a property of systems, it was found that two systems may have the same energy, yet one is capable of performing work and one is not. That is to say: not all energy is the same. Put another way, some forms of energy (or states of a system) are more capable (or useful) than others. This led to the concept of *available* and *unavailable* energy—more precisely *free energy*—that is, energy that is free to perform work and that which is not.

---

[1] Ludwig Eduard Boltzmann (1844–1906), Austrian physicist and philosopher, father of statistical mechanics.

[2] $k=1.38065\times10^{-23}$ J/K is a physical constant, which relates the average relative kinetic energy of particles in a gas with the temperature and also appears in Planck's law of black-body radiation.

When considering thermodynamic systems, the distinction between *open* and *closed* is very important. An *open* system exchanges mass (in, out, or both) with the surroundings (i.e., everything that isn't the system); whereas, *closed* systems do not. For reasons we have discussed elsewhere, the appropriate specific energy (i.e., per unit mass) for a closed system is called *internal energy* and given the symbol $u$. For an open system this quantity is enthalpy and given the symbol $h$. The free energy for a closed system is given by the following expression named after Helmholtz.[3]

$$a = u - Ts \qquad (1.2)$$

The free energy for an open system is given by the following expression named after Gibbs.[4]

$$g = h - Ts \qquad (1.3)$$

We have the Helmholtz Free Energy (HFE) and the Gibbs Free Energy (GFE). As both the HFE and GFE contain the term, $-Ts$, a process which proceeds toward the maximum entropy would also be in the direction of minimal free energy. Thus we have the guiding principle of thermochemical reactions:

### *Free Energy Minimization*

The internal energy or enthalpy of reaction products may change depending on heat transfer into or out of a system, but the entropy will always be increasing, which is the Second Law of Thermodynamics. Not all reaction outcomes (i.e., products) are possible. For instance, there is a fixed number of each atom in the reactants and we must also conserve energy (i.e., the First Law of Thermodynamics). We must, therefore, limit (or constrain) the products to only those, which are possible. The relationships between energy, enthalpy, entropy, and free energy are nonlinear. The mathematical process we will use to solve thermochemical reactions is:

### *Nonlinear Constrained Minimization*

---

[3] Hermann Ludwig Ferdinand von Helmholtz (1821–1894) German physician and physicist who made significant contributions in several scientific fields.

[4] Josiah Willard Gibbs (1839–1903) American scientist who made significant theoretical contributions to physics, chemistry, and mathematics.

# Chapter 2. Simple Dissociation

We will begin with the simplest reaction: dissociation of a diatomic gas (e.g., hydrogen, nitrogen, oxygen, fluorine, chlorine, bromine, or iodine). We will also restrict our preliminary discussion to a *perfect* gas, that is, an *ideal* gas having constant specific heats. An ideal gas follows the ideal gas law:

$$p = \rho R T \tag{2.1}$$

The internal energy, u, and enthalpy, h, of this perfect gas can be expressed:

$$u = u_F + C_V T$$
$$h = h_F + C_p T \tag{2.2}$$

where $C_P$ and $C_V$ are the constant pressure and volume specific heats, respectively. The terms $u_F$ and $h_F$ account for the fact that there is some internal energy and enthalpy associated with the *formation* of the molecules, that is, arising from the molecular structure. We recognize that $H_2O$ is not simply two hydrogens *associating* with each oxygen, but rather a different molecule, having a specific structure. The entropy of this perfect gas can be expressed:

$$s = s_F + C_p \ln T - R \ln P \tag{2.3}$$

where $s_F$ is the entropy of formation. The Helmholtz and Gibbs free energies for this perfect gas can then be expressed:

$$a = u_F + C_V T - T\left(s_F + C_p \ln T - R \ln P\right)$$
$$g = h_F + C_p T - T\left(s_F + C_p \ln T - R \ln P\right) \tag{2.4}$$

We will defer explanation of how these properties (R, $C_P$, $C_V$, $u_F$, $h_F$, and $s_F$) are obtained. Suffice it to say that these can be found online or in textbooks, including CRC[5], Lange[6], and Perry & Chilton.[7] We will also postpone discussion of reference and ground states, as well as structural entropy at absolute zero, as these are not essential for our introductory examples.

We also need to introduce Dalton's Partial Pressure Law,[8] which states that (at least for an ideal mixture of ideal gases), each constituent contributes a *partial* pressure (i.e., only a *part* of the *total*). The sum of these partial pressures is equal to the total pressure. Each partial pressure is equal to the total pressure

---

[5] *CRC Handbook of Chemistry and Physics*, (99 editions) published by CRC Press, Boca Raton, Florida.

[6] Lange, *Handbook of Chemistry*, (multiple editions) published by McGraw-Hill, New York.

[7] Perry and Chilton, *Chemical Engineer's Handbook*, (multiple editions) published by McGraw-Hill, New York.

[8] John Dalton (1766–1844) English chemist, physicist, and meteorologist, best known for introducing the atomic theory into chemistry.

times the mole fraction of that constituent. The partial pressure ratios (i.e., individual partial pressures divided by the total pressure) are equal to the mole fractions, which also sum to unity.

With these definitions in hand, we can now consider our first dissociation reaction, that of diatomic oxygen:

$$O_2 \Leftrightarrow (1-\delta)O_2 + 2\delta O \qquad (2.5)$$

Numerical values for these and other substances can be found in spreadsheet gases.xls in the examples\dissociation folder. Implementation of this particular equation can be found in spreadsheet dissociation_of_oxygen.xls. The following figure shows the Gibbs free energy, $g$, over a range of temperatures and dissociation fractions:

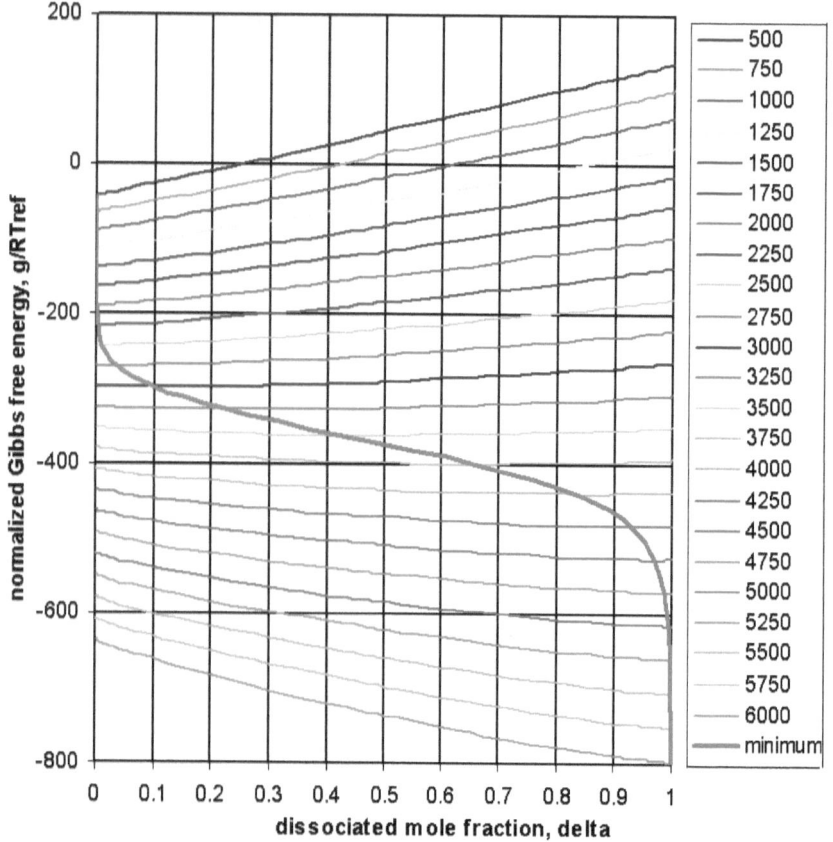

Figure 5. Locus of Free Energy Minima

Each of the thin curves (500°K to 6000°K) going from left to right illustrates the impact of $\delta$ on $g$. The thick red curves runs through the lowest point in each

4

of the other curves that exhibit a minimum. This represents the equilibrium dissociation of oxygen over this range of temperatures. Note that the red curve asymptotically approaches zero on the left and one on the right. This means that at low temperatures there is no dissociation and at high temperatures there is complete dissociation. The calculation is implemented inside the spreadsheet with VBA® macros. The same is available in C (O2diss.c) in the same folder:

```
double Pref=0.101325;   /* MPa */
double Tref=25.+273.15;/* K */
double R    =8.3144621; /* kJ/kg-mole/K */

typedef struct{double Hf,Sf,Cp;}HSC; /* kJ/kg/K */
HSC O2={       0.,205.2,28.01};
HSC O ={259336.,161.1,21.93};

double G(HSC hsc,double T,double P)
  {
  return((hsc.Hf+hsc.Cp*(T-Tref)-
    T*(hsc.Sf+hsc.Cp*log(T/Tref)-R*log(P/Pref)))/R/Tref);
  }

double Gibbs(double T,double P,double delta)
  {
  return((1.-delta)*G(O2,T,P*(1.-delta)/(1.+delta))
    +2.*delta*G(O,T,P*2.*delta/(1.+delta)));
  }

double dGd(double T,double P,double delta)
  {
  double d1,d2,g1,g2;
  if(delta<0.5)
    {
    d1=delta;
    d2=delta+0.00001;
    }
  else
    {
    d1=delta-0.00001;
    d2=delta;
    }
  g1=Gibbs(T,P,d1);
  g2=Gibbs(T,P,d2);
  return((g2-g1)/(d2-d1));
  }

double dmin(double T,double P)
  {
  int iter;
  double d1,d2,delta;
```

```
  d1=0.;
  d2=1.;
  for(iter=0;iter<64;iter++)
     {
     delta=(d1+d2)/2.;
     if(dGd(T,P,delta)<0.)
        d1=delta;
     else
        d2=delta;
     }
  return(delta);
  }

int main(int argc,char**argv,char**envp)
  {
  double delta,g,T;
  printf("T,delta,g\n");
  for(T=500.;T<6001.;T+=250.)
     {
     delta=dmin(T,Pref);
     g=Gibbs(T,Pref,delta);
     printf("%.0lf,%lf,%.1lf\n",T,delta,g);
     }
  return(0);
  }
```

The output of the program above is:

```
T,delta,g
500,0.000000,-42.0
750,0.000000,-64.8
1000,0.000000,-88.5
1250,0.000000,-113.0
1500,0.000000,-138.0
1750,0.000020,-163.4
2000,0.000256,-189.3
2250,0.001623,-215.6
2500,0.007105,-242.2
2750,0.023945,-269.4
3000,0.066253,-297.4
3250,0.156353,-326.8
3500,0.317731,-358.7
3750,0.541900,-394.0
4000,0.753962,-433.0
4250,0.886490,-474.8
4500,0.949623,-518.2
4750,0.976983,-562.4
5000,0.988913,-607.1
5250,0.994349,-652.1
5500,0.996962,-697.3
5750,0.998286,-742.8
```

6000,0.998991,-788.4

A finite difference is used to calculate the slope of the free energy curve and a bisection search is used to locate the minimum, where the slope is zero. By simply changing constants, we can analyze the dissociation of diatomic nitrogen. Air contains several constituents, including moisture (i.e., water vapor). While the process of locating the minimum free energy with several variables is more complicated, it is conceptually the same so that we can generate the following plot using the program described in Appendices A and B:

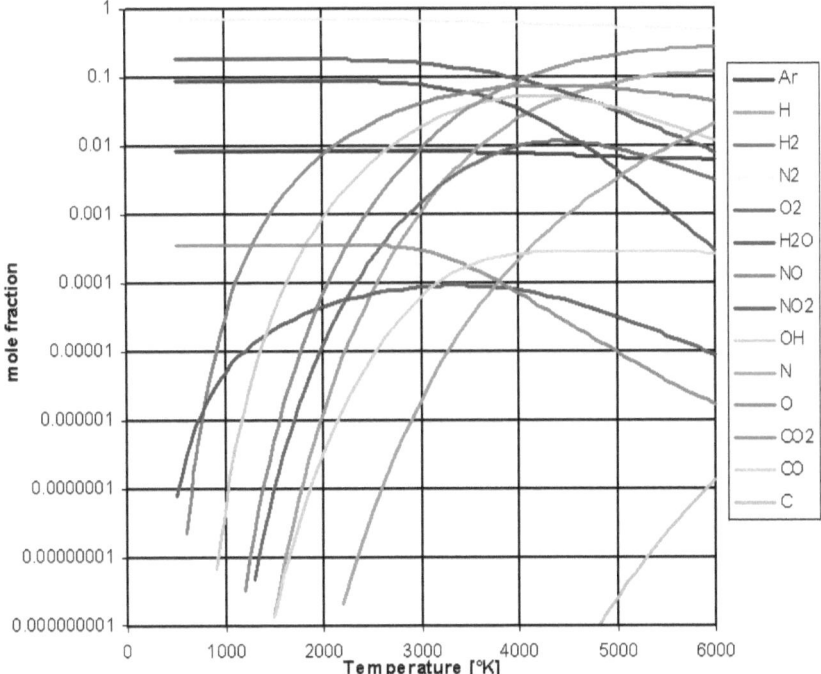

**Figure 6. Dissociation of Air**

As shown in this figure, there are a variety of dissociated species, even in the absence of what is often considered a simple reaction. There is even some dissociated carbon at very high temperatures (upward diagonal light magenta curve in the lower right corner of the above graph). Even this variety is still *simple* dissociation because there are important factors we have yet to consider, including non-ideal behavior. We assumed an ideal mixture of ideal gases, which is not the way things work in the real world. Real molecules interact differently with each other and even individually depart from $P=\rho RT$. We will cover these details in subsequent chapters

7

# Chapter 3. Simple Combustion

Without explicitly stating it, we introduced the most important factor in combustion with Equation 2.4, namely, $h_F$, the enthalpy (or heat) of formation. Buried inside the spreadsheet dissociation_of_oxygen.xls at the top of the VBA code are the following statements:

```
Const CpO As Double = 21.926 'kJ/kg-mole/C
Const CpO2 As Double = 28.01 'kJ/kg-mole/C
Const HfO As Double = 249336 'kJ/kg-mole
Const HfO2 As Double = 0
Const Pref As Double = 0.101325 'MPa
Const R As Double = 8.3144621 'kJ/kg-mole/C
Const SfO As Double = 161.07 'kJ/kg-mole/C
Const SfO2 As Double = 205.15 'kJ/kg-mole/C
Const Tref As Double = 25 + 273.15 'K
```

The heat of formation of O is 249,335 kJ/kg-mole and the heat of formation of $O_2$ is zero. This means that if you want to split one kg-mole $O_2$ into 2 kg-moles of O, it will *require* 2x249,335=498,670 kJ to accomplish. The heat of formation of diatomic $O_2$ is zero because this is the state in which oxygen naturally occurs. Split (non-diatomic) oxygen is unnatural. The two atoms must be pulled apart. An $O_2$ molecule is diatomic in the same way that the Moon is *bound* to the Earth—its current total energy level (kinetic + gravitational potential) is less than it would be if the Moon were free of the Earth. You would have to *add energy* to the Moon to get it away from the Earth.

## Reference Conditions and Units

The reference conditions are of great importance. You will find these explicitly (or perhaps implicitly) stated along with any table of properties, such as the one used to create the preceding VBA statements. I can't stress enough that if you change $P_{REF}$ or $T_{REF}$ then $h_F$ and $s_F$ will also change. Failure to recognize this fact has led to countless arguments and much wasted effort. One example (of which I am personally aware) is the ASME-PTC22 committee discussion over the gas turbine energy balance at different reference conditions. I will discuss (and provide) a spreadsheet showing exactly how $h_F$ changes when you change $T_{REF}$.

It is also important when using the SI system of units to distinguish between a gram-mole and a kilogram-mole, as these are not the same. Of course, there is only one type of pound-mole, so three orders of magnitude errors don't slip in as easily. I also caution you regarding the use of tabular data, especially in SI units, as publishers (and internet posters) are often quite loose and sloppy when it comes to units. It helps to know the magnitude of the numbers when you are looking for them online. Consider the following table, which I found at the top of a Google™ search:

9

| substance | $M/(g\,mol^{-1})$ | $\Delta_f H^{\circ}/(kJ\,mol^{-1})$ | $\Delta_f G^{\circ}/(kJ\,mol^{-1})$ | $S_m^{\circ}/(J\,K^{-1}\,mol^{-1})$ | $C_{p,m}^{\circ}/(J\,K^{-1}\,mol^{-1})$ |
|---|---|---|---|---|---|
| $H_2S(g)$ | 34.08 | −20.63 | −33.56 | 205.79 | 34.23 |
| $Sn(s, \beta)$ | 118.69 | 0 | 0 | 51.55 | 26.99 |
| $Sn(g)$ | 118.69 | +302.1 | +267.3 | 168.49 | 20.26 |
| $Zn(s)$ | 65.37 | 0 | 0 | 41.63 | 25.40 |
| $Zn(g)$ | 65.37 | +130.73 | +95.14 | 160.98 | 20.79 |
| $Zn^{2+}(aq)$ | 65.37 | −153.89 | −147.06 | −112.1 | 46 |
| $ZnO(s)$ | 81.37 | −348.28 | −318.30 | 43.64 | 40.25 |

I have circled the mixture of kJ and J—very sloppy. If you're going to use kilojoules, you should also use kilograms, including kg-moles and not g-moles. If you don't get this right, you may end up with disastrous results, like missing Mars and squandering a hundred million dollars.[9] Such blunders occur when people use physical quantities (units) with no comprehension of what they mean. One way to minimize such errors is to always divide $a$, $g$, $h$, and $u$ by $RT_{REF}$ and $C_P$ by $R$. In this way you can become accustomed to the expected magnitudes.

### Endothermic vs. Exothermic Reactions

Endothermic reactions absorb heat, while exothermic reactions release heat. Splitting diatomic oxygen into two atoms is *endo*thermic, that is, it requires energy to accomplish. Burning methane (i.e., reacting it with oxygen) is *exo*thermic, producing heat. Energy is neither created nor lost, which is the First Law of Thermodynamics. Energy changes from one form to another. Methane contains energy that came from the sun. Plants take in energy, carbon dioxide, and water. Plants build structures of hydrocarbons and return oxygen to the atmosphere. Combustion of hydrocarbons, such as methane, releases the chemical energy that was stored in molecular structure. This is the heat of formation, $h_F$.

### Complete Combustion

We will first consider simple complete combustion of diatomic hydrogen with diatomic oxygen, producing water.

$$2H_2 + O_2 \rightarrow 2H_2O \tag{3.1}$$

This equation is what you typically get in chemistry—balancing the atoms on each side, assuming a fixed outcome (i.e., the *products* of the reaction). Given these assumptions, we need not consider minimizing free energy, but

---

[9] In October 1999 NASA lost a $125-million Mars Climate Orbiter because spacecraft engineers failed to convert from English to metric measurements when exchanging vital data before the craft was launched.

merely the conservation of energy. If the specific heats are presumed constant, the solution is straightforward.

$$2\left(h_{FH_2} + C_{PH_2}\left(T_1 - T_{REF}\right)\right)$$
$$+ \left(h_{FO_2} + C_{PO_2}\left(T_1 - T_{REF}\right)\right) \tag{3.2}$$
$$= 2\left(h_{FH_2O} + C_{PH_2O}\left(T_2 - T_{REF}\right)\right)$$

The heat of formation of diatomic hydrogen ($h_{FH_2}$) and diatomic oxygen ($h_{FO_2}$) are both zero. The heat of formation for water ($h_{FH_2O}$) is −241,981 kJ/kg-mole. The specific heats for diatomic hydrogen ($C_{PH_2}$), diatomic oxygen ($C_{PO_2}$), and water ($C_{PH_2O}$) are 27.33, 28.01, and 28.77 kJ/kg-mole/°K, respectively. The initial temperature ($T_1$) is known. If we set $T_1 = T_{REF}$, Equation 3.2 reduces to:

$$T_2 = T_{REF} - \frac{h_{FH_2O}}{C_{PH_2O}} = 8709°K \tag{3.3}$$

Considering the melting point of tungsten 3695°K, you can see why a hydrogen torch can cut through almost anything and also why they are so dangerous. The actual flame temperature will be much less, because there is significant dissociation under such conditions, as illustrated in the last graph of the previous chapter.

<u>Adiabatic and Isobaric Processes</u>

In solving this combustion equation, we assumed there was no heat transfer into or out of the combustor. Such is called an *adiabatic* process (i.e., Q=0). We also implicitly presumed that combustion at a constant pressure of one atmosphere. Such is called an *isobaric* process (i.e., ΔP=0). If we allow for dissociation, including: H, O, and OH, and adjust all of the mole fractions to conserve the number of hydrogen and oxygen atoms and also minimize the free energy, we arrive at the following result:

```
T2=3493.3 K,  P=0.101325 MPa,  Q=0 kJ
substance      y          x        h/RTo   s/R    g/RTo
------------------------------------------------------------
Hdia        2.0000000  0.6666667    2.8   15.46   -25.0
Odia        1.0000000  0.3333333    3.0   25.24   -42.5
------------------------------------------------------------
reactants   3.0000000  1.0000000    8.5   56.15   -92.6
============================================================
H           0.1346769  0.0567545  114.7   20.12  -120.9
Hdia        0.3372400  0.1421171   43.3   24.35  -241.8
O           0.0712162  0.0300114  128.7   26.66  -183.5
Odia        0.0989444  0.0416964   47.6   35.41  -367.1
Water       1.4599482  0.6152402  -35.3   33.31  -425.3
Hydroxl     0.2709468  0.1141803   59.4   31.07  -304.4
------------------------------------------------------------
```

11

```
products    2.3729725 1.0000000   8.5 73.38 -850.7
----------------------------------------------------
difference -0.6270275 0.0000000   0.0 17.22 -758.1
```

There is no heat transfer, as indicated by 0.0 at the bottom of the h/RTo column. The initial and final normalized enthalpies are both 8.5 [unitless]. This process has increased the normalized entropy, s/R, by 17.22 [unitless]. The reduced free energy, g/RT, has been reduced by 758.1 [unitless]. The final temperature, T2, is 3493°K, less than half the 8439°K, obtained by neglecting dissociation.

*This is one of those,*
*"Oh, yeah, we didn't tell you about this in undergraduate*
*school because you wouldn't understand," moments.*
*But wait... it gets worse...*
*We haven't even discussed variable specific heats and non-*
*ideal behavior of the gases yet, nor have we explained how to*
*solve this nest of simultaneous nonlinear equations.*
*Be patient...*
*we'll get there eventually.*

The y-column lists the number of moles. The x-column lists the mass fractions. In the y-column we see that, instead of forming 2 moles of $H_2O$, we form only 1.46 moles, so that's where some of our seemingly lost heat of formation went. The rest went into dissociating the H, O, and OH.

### Incomplete Combustion

The preceding reaction is also *complete* combustion, in that there was sufficient oxygen to combine with all the hydrogen, if it would happen that way, which it won't. If there were insufficient oxygen, we would necessarily have *incomplete* combustion. This time we will provide only half the required oxygen, keeping everything else the same.

```
T2=3141.3 K, P=0.101325 MPa, Q=0 kJ
substance       y         x      h/RTo  s/R    g/RTo
----------------------------------------------------
Hdia      2.0000000 0.8000000   2.8 15.27  -24.7
Odia      0.5000000 0.2000000   3.0 25.75  -43.4
----------------------------------------------------
reactants 2.5000000 1.0000000   7.0 43.42  -71.1
====================================================
H         0.0877855 0.0425607 111.7 20.14 -100.4
Hdia      0.9754445 0.4729209  37.9 22.66 -200.7
O         0.0019510 0.0009459 125.6 29.84 -188.6
Odia      0.0006347 0.0003077  41.8 39.80 -377.3
Water     0.9645460 0.4676371 -43.3 32.87 -389.3
Hydroxl   0.0322335 0.0156276  54.2 32.59 -289.0
----------------------------------------------------
```

```
products    2.0625953 1.0000000    7.0 56.70 -590.1
---------------------------------------------------
difference -0.4374047 0.0000000    0.0 13.28 -518.9
```

Only 0.9645 moles of $H_2O$ are produced and the final temperature is now only 3141°K. The total enthalpy remains the same before and after (Q=0), the normalized entropy increases by 13.28, and the normalized free energy is reduced by 518.9. Overall, there are fewer moles of the dissociated components.

## Excess Oxygen/Air

Most often we prefer to provide more than enough oxygen, usually in the form of air. In this case we call the over-abundance of oxidizing gas, *excess air*. Before considering the constituents of air, we will first double the oxygen and solve the hydrogen combustion equation again.

```
T2=3240.4K, P=0.101325 MPa, Q=0 kJ
substance        y         x      h/RTo   s/R    g/RTo
------------------------------------------------------
Hdia       2.0000000 0.5000000    2.8   15.74   -25.6
Odia       2.0000000 0.5000000    3.0   24.83   -41.7
------------------------------------------------------
reactants  4.0000000 1.0000000   11.5   81.15  -134.6
======================================================
H          0.0405755 0.0125982  112.5   21.43  -120.3
Hdia       0.0778819 0.0241813   39.4   25.77  -240.6
O          0.1242465 0.0385769  126.5   26.21  -158.3
Odia       0.8977103 0.2787273   43.4   33.14  -316.6
Water      1.7233276 0.5350707  -41.0   32.94  -398.8
Hydroxl    0.3570054 0.1108455   55.6   30.77  -278.6
------------------------------------------------------
products   3.2207472 1.0000000   11.5  103.64 -1114.2
------------------------------------------------------
difference -0.7792528 0.0000000    0.0   22.48  -979.6
```

The final temperature is now 3240°K. More moles of $H_2O$ are created than either of the two preceding cases (1.7233), so more of the heat of formation has been recovered, but some of this went into heating up the excess oxygen. This third case also has the greatest increase in normalized entropy (22.48) and greatest decrease in normalized free energy (979.6).

*This is an application of the "no free lunch"*
*(or "you don't get something for nothing") principle.*

## Combustion of Hydrocarbons

When we add carbon the number of products goes up, as does the number of equations and the difficulty of solving them simultaneously. To recap, we are holding the pressure constant ($\Delta P=0$, isobaric) and keeping all of the heat inside the combustor ($\Delta h=Q=0$, adiabatic). We must conserve the elements (i.e., same number of C, O, and H in the reactants and products). We must also minimize the Gibbs free energy with respect to the number of moles of each product species ($\partial g/\partial y=0$). We minimize the Gibbs free energy because this is an isobaric process. If this were an *isochoric* (i.e., constant-volume) process, we would minimize the Helmholtz free energy. We are simultaneously solving the First and Second Laws of Thermodynamics. We are still assuming perfect gases, that is constant specific heats and $P=\rho RT$.

We will first consider combustion of methane ($CH_4$) with sufficient diatomic oxygen and no disassociation:

$$CH_4 + 2O_2 \rightarrow CO2 + 2H_2O \qquad (3.4)$$

Substituting in the heats of formation and specific heats in SI units from the list in spreadsheet gases.xls in the examples folder:

$$
\begin{aligned}
&(-74902 + 20.55(T_1 - T_{REF})) \\
&+ 2(0 + 28.01(T_1 - T_{REF})) \\
&= (-393769 + 33.57(T_2 - T_{REF})) \\
&+ 2(-241981 + 28.77(T_2 - T_{REF}))
\end{aligned}
\qquad (3.5)
$$

If $T_1=T_{REF}=298.15°K$, the solution is $T_2=9110°K$, which is much higher than actual flame temperatures for burning methane, even in pure oxygen for the same reason as before.

### Combustion of Methane with Dissociation

We next add the other products, which include dissociations. The solution is obtained as before with the program described in Appendices A and B:

```
T2=3410.5 K,  P=0.101325 MPa,  Q=0 kJ
substance        y           x         h/RTo    s/R      g/RTo
---------------------------------------------------------------
Methane     1.0000000  0.3333333    -26.0    32.11    -83.8
Odia        2.0000000  0.6666667      3.0    24.54    -41.2
---------------------------------------------------------------
reactants   3.0000000  1.0000000    -20.1    81.20   -166.2
===============================================================
H           0.1086816  0.0298157    114.0    20.70   -122.7
Hdia        0.2103005  0.0576938     42.0    25.14   -245.4
O           0.1217421  0.0333987    128.0    26.49   -174.9
Odia        0.2908651  0.0797959     46.2    34.64   -349.8
Water       1.5571946  0.4272007    -37.2    33.51   -420.3
Hydroxl     0.3563282  0.0977551     58.2    31.12   -297.6
C           0.0000000  0.0000000    315.1    45.08   -200.2
Cmonox      0.6169950  0.1692664      0.1    32.83   -375.1
Cdiox       0.3830050  0.1050736    -86.8    40.52   -550.1
---------------------------------------------------------------
```

```
products    3.6451121 1.0000000 -20.1 119.89 -1390.6
-----------------------------------------------------
difference 0.6451121 0.0000000   0.0  38.68 -1224.4
```

The final temperature is 3410.5°K, much less than the ideal 9110°K. The change in normalized enthalpy ($\Delta h/RTo$) is 0.0, the increase in normalized entropy ($\Delta s/R$) is 38.68 (always >0), and the decrease in normalized free energy ($\Delta g/RTo$) is 1224.8 (always <0). We only form 0.383 moles of $CO_2$ (instead of the ideal 1.000) and only 1.557 moles of $H_2O$ (instead of the ideal 2.000).

### Incomplete Combustion of Methane with Dissociation

As with the combustion of hydrogen, we can reduce the available oxygen by one-half and repeat the process to obtain:

```
T2=2805.3 K, P=0.101325 MPa, Q=0 kJ
substance       y         x       h/RTo   s/R    g/RTo
------------------------------------------------------
Methane     1.0000000 0.5000000  -26.0  31.71   -83.1
Odia        1.0000000 0.5000000    3.0  24.83   -41.7
------------------------------------------------------
reactants   2.0000000 1.0000000  -23.1  56.54  -124.8
======================================================
H           0.0387098 0.0127993  108.9  21.06   -89.1
Hdia        1.0637633 0.3517318   32.9  22.45  -178.3
O           0.0002367 0.0000783  122.6  32.03  -178.6
Odia        0.0000688 0.0000227   36.4  41.86  -357.2
Water       0.9121829 0.3016120  -50.8  32.55  -356.9
Hydroxl     0.0093978 0.0031074   49.2  33.71  -267.7
C           0.0000000 0.0000000  310.0  46.29  -125.3
Cmonox      0.9219550 0.3048431   -9.1  31.35  -303.9
Cdiox       0.0780450 0.0258055 -102.2  40.44  -482.5
------------------------------------------------------
products    3.0243593 1.0000000  -23.1  86.77  -839.0
------------------------------------------------------
difference  1.0243593 0.0000000   -0.0  30.23  -714.2
```

The final temperature drops from 3410.5°K to 2805.3°K. Only 0.0780 moles of $CO_2$ and 0.9122 moles of $H_2O$ are formed. *This is a significant result* and arises from the fact that $h_{FCO2}$=-208,586 while $h_{FH2O}$=-412,735 kJ/kg-mole, almost twice as large.

*When oxygen-limited, combustion of a*
*hydrocarbon will prefer H2O over CO2.*

Notice also that only 0.0094 moles of OH are produced, while 0.9222 moles of CO are produced, almost twelve times the number of moles of $CO_2$. This is one of the many reasons why excess air is recommended for complete combustion and reduced emissions.

15

## Combustion of Methane with Excess Oxygen

We double the ideal moles of oxygen and solve to obtain:

```
T2=3142.0 K, P=0.101325 MPa, Q=0 kJ
```

| substance | y | x | h/RTo | s/R | g/RTo |
|-----------|-----------|-----------|-------|--------|--------|
| Methane | 1.0000000 | 0.2000000 | -26.0 | 32.63 | -84.7 |
| Odia | 4.0000000 | 0.8000000 | 3.0 | 24.36 | -40.9 |
| reactants | 5.0000000 | 1.0000000 | -14.2 | 130.08 | -248.3 |
| H | 0.0323402 | 0.0060296 | 111.7 | 22.09 | -121.0 |
| Hdia | 0.0507310 | 0.0094584 | 37.9 | 26.57 | -242.0 |
| O | 0.1755122 | 0.0327229 | 125.6 | 26.30 | -151.4 |
| Odia | 1.9672332 | 0.3667753 | 41.8 | 32.72 | -302.8 |
| Water | 1.7284221 | 0.3222508 | -43.3 | 33.24 | -393.3 |
| Hydroxl | 0.4093535 | 0.0763208 | 54.2 | 31.00 | -272.4 |
| C | 0.0000000 | 0.0000000 | 312.9 | 49.46 | -208.1 |
| Cmonox | 0.2477541 | 0.0461918 | -4.0 | 33.75 | -359.4 |
| Cdiox | 0.7522459 | 0.1402504 | -93.6 | 39.61 | -510.8 |
| products | 5.3635922 | 1.0000000 | -14.2 | 179.35 | -1903.0 |
| difference | 0.3635922 | 0.0000000 | 0.0 | 49.27 | -1654.7 |

The final temperature only drops from 3410.5°K to 3142.0°K. Eleven percent more $H2O$ (1.7284 moles) and almost twice as much (+96%) $CO2$ (0.7522 moles) is produced by doubling the available oxygen. This is even more incentive to provide excess air for combustion. This third case also has the largest increase in entropy and largest decrease in free energy.

16

## Chapter 4. Mathematical Formulation

Before we consider more complex reactions and more accurate properties, including variable specific heats and non-ideal gas behavior, we must present the underlying equations and how they can be solved. The reactants will be indicated by the subscript, R, and products, P. The moles of each species (i.e., distinct atom or molecule) will be given the symbol, y. Totals will be indicated by the summation sign, $\Sigma$. We will develop the equations using Gibbs free energy, g, which applies to isobaric processes, recognizing that Helmholtz free energy, a, is used for isochoric processes.

### Three-Part Problem

We must solve three requirements, which will expand to more than three equations: 1) conservation of elements (e.g., the same number of C, H, and O on either side of the reaction), 2) conservation of energy (i.e., the First Law of Thermodynamics), 3) minimum free energy (i.e., the Second Law of Thermodynamics).

$$\sum \langle y_{EL} \rangle_R = \sum \langle y_{EL} \rangle_P \quad EL = C, H, O, \ldots \quad (4.1)$$

$$\sum \langle y_I h_I \rangle_R + Q = \sum \langle y_I h_I \rangle_P \quad (4.2)$$

$$\frac{\partial \sum \langle y_I g_I \rangle}{\partial y_I} = 0 \quad (4.3)$$

Equation 4.1 expands to the number of distinct elements, $n_E$, (e.g., C, H, O = 3). Equation 4.2 does not expand. Equation 4.3 expands to the number of products, $n_P$. The total number of equations is then $n_E + 1 + n_P$. Equations 4.1 and 4.2 are linear, while Equation 4.3 is nonlinear. Consider the simplest expression from Equation 2.4, which presumes ideal gas behavior (including Dalton's Partial Pressure Law) and constant specific heat:

$$g_I = h_{FI} + C_{PI}(T - T_{REF}) - T\left[ s_{FI} + C_{PI} \ln\left(\frac{T}{T_{REF}}\right) - R \ln\left(\frac{P_I}{P_{REF}}\right) \right] \quad (4.4)$$

$$P_I = \frac{y_I}{\sum y_I} P$$

$$\frac{\partial(y_I g_I)}{\partial y_I} = g_I + y_I \frac{\partial g_I}{\partial y_I} \quad (4.5)$$

$$\frac{\partial g_I}{\partial y_I} = \frac{RT}{y_I} \quad (4.6)$$

$$\frac{\partial(y_I g_I)}{\partial y_I} = g_I + RT \quad (4.7)$$

17

Equation 4.4b is Dalton's Partial Pressure Law. Note the limit as $y_I \to 0$:

$$\lim_{y_I \to 0} \left\{ y_I \ln\left[ \left( \frac{y_I}{\sum y_I} \right) \left( \frac{P}{P_{REF}} \right) \right] \right\} = 0 \tag{4.8}$$

This equation is meaningful because the limit is defined, but it is problematic to implement in code, as the limit does not evaluate with software. It is, therefore, necessary to restrict the lower bound on all mole fractions to some value, such as FLT_EPSILON (1.19E-7) or DBL_EPSILON (2.22E-16), which are defined in float.h.

## Nonlinear Constrained Minimization

Apart from the aforementioned restrictions, Equations 4.1 and 4.2 constitute a set of simultaneous linear equations. Equation 4.3 is a set of nonlinear equations, consisting of Equations 4.4 through 4.8 for each product. The most effective way of solving this combination is Lagrange multipliers, which are covered in any advanced calculus text. Finding the moles of each product that result in the minimum free energy while conserving the number of atoms as well as energy is the classic constrained nonlinear minimization problem. Most algorithms for accomplishing this either use Newton's method or are based on it.

The temperature of the products controls the conservation of energy and is required to calculate each of the partial derivatives of the free energy. It is, therefore, different in that respect. Rather than also calculating all of the partial derivatives with respect to the temperature of the products, which would be necessary to include it in the Newton iteration, it is more effective to assume a value, calculate each step of the minimization problem, and update outside the Newton iteration, as illustrated below:

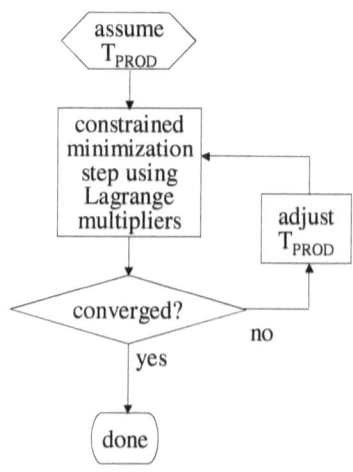

## Lagrange Multipliers

While the topic of Lagrange multipliers is presented in many texts on advanced calculus, the implementation is often omitted or vague; therefore, we will address the subject here. The minimization problem, typically implemented as a Newton iteration for Equation 4.3, can be expressed in matrix form:

$$[A][y] = [B] \qquad (4.9)$$

where A is a matrix of dimension $n_Px n_P$, y is a column matrix of dimension $n_Px1$, and B is also a column matrix of dimension $n_Px1$. Matrix A contains the second partial derivatives of the total free energy with respect to the product moles:

$$A_{I,J} = \frac{\partial^2 G}{\partial y_I \partial y_J} = \begin{cases} \left[ \dfrac{RT\left(\sum y_I - y_J\right)}{y_J \sum y_I} \right]_{I=J} \\ \left[ \dfrac{-RT}{\sum y_I} \right]_{I \neq J} \end{cases} \qquad (4.10)$$

The matrix B contains the negative of the first partial derivatives of the total free energy with respect to the product moles:

$$B_I = -\frac{\partial G}{\partial y_I} = -g_I \qquad (4.11)$$

The conservation of elements (Equation 4.1) can also be expressed in matrix form:

$$[C][y] = [D] \qquad (4.12)$$

where C is a matrix of dimension $n_Ex n_P$, y is the same $n_Px1$, and D is a column matrix of dimension $n_Ex1$. Consider the conservation of oxygen atoms in Equation 3.4. There are two molecules of diatomic oxygen (2x2=4) on the left, and on the right an unknown number of molecules of carbon dioxide ($y_{CO2}x2$) and water ($y_{H2O}$). The oxygen row of the matrices in Equation 4.12 would then be:

$$[2,2]\begin{bmatrix} y_{CO2} \\ y_{H2O} \end{bmatrix} = [4] \qquad (4.13)$$

In the simple combustion of methane (Equation 3.4), there will be three rows in Equation 4.12, one each for oxygen, hydrogen, and carbon. The step that is most often left out of textbooks is how Equations 4.9 and 4.12 are combined to form a single matrix that can be solved using any one of the various techniques. The result is:

$$
\begin{array}{|c|c|}
\hline
A & C^T \\
\hline
C & 0 \\
\hline
\end{array}
\; x \;
\begin{array}{|c|}
\hline
y \\
\hline
\Lambda \\
\hline
\end{array}
=
\begin{array}{|c|}
\hline
B \\
\hline
D \\
\hline
\end{array}
\tag{4.14}
$$

The first part $[AC^TC0]$ is square, having dimensions $(n_P+n_E) \times (n_P \times n_E)$. Matrices A, B, C, and D are as before and $C^T$ is the transpose of C. Column vectors $[y\Lambda]$ and $[BD]$ both have dimension $(n_P+n_E) \times 1$. The new portion, $\Lambda$, is a column vector of dimension $n_E \times 1$. These are the Lagrange multipliers. Their numerical value is not immediately apparent, but is of consequence, as we will discuss later.

It is important to note that $[A][y]$ no longer exactly equals $[B]$, because there are now additional terms $[C^T][\Lambda]$ that contribute to each row. If you insist that $[C][\Lambda]=[D]$, then $[A][y]$ *can't* be exactly $[B]$. This is the *cost* of requiring that the elemental abundances be conserved, which is non-negotiable. Minimizing free energy (i.e., the Ay=B part) is negotiable, in that not all solutions are possible. Another way to consider this is to conceptualize all product molar quantities as a landscape. Some sections of the landscape are not accessible. This is the *constrained* part of minimization process. Conceptually, the accessible part of the landscape is inside a fence, which is an expression of the elemental abundances.

### Implementation

Implementing these equations for general problems is quite complex and includes interpreting the reaction and parsing the compounds. Before we tackle that challenge, we will implement the simplest case covered so far, combustion of hydrogen with dissociation:

$$2H_2 + O_2 \rightarrow H + H_2 + O + O_2 + H_2O + OH \tag{4.13}$$

There are only two elements ($n_E=2$), H and O. There are six unknown product moles ($n_P=6$), $y_H$, $y_{H2}$, $y_O$, $y_{O2}$, $y_{H2O}$, and $y_{OH}$. All of the properties ($h_F$, $s_F$, and $C_P$) can be found in gases.xls. The code (H2comb.c) can be found in the folder examples\combustion. There will be seven unknowns (six moles plus one temperature) plus three Lagrange multipliers, for a total of ten simultaneous equations. We will solve nine equations using Gauss elimination and one (temperature of the products) separately in the outer iteration loop. Ions (e.g., $OH^-$, $e^-$, etc.) are present in this reaction, but we will postpone covering these until later. The property calculations are simple:

```
double h(PROP prop,double T) /* enthalpy of one product
    */
{
return(prop.Hf+prop.Cp*(T-Tref));
```

20

```
    }

double hR() /* enthalpy of all reactants combined */
    {
    int i;
    double hr;
    for(hr=i=0;i<nR;i++)
      hr+=yR[i]*h(pR[i],T1);
    return(hr);
    }

double hP() /* enthalpy of all products combined */
    {
    int i;
    double hp;
    for(hp=i=0;i<nP;i++)
      hp+=yP[i]*h(pP[i],T2);
    return(hp);
    }

double g(PROP prop,double T,double P) /* Gibbs free
    energy of one product */
    {
    return(prop.Hf+prop.Cp*(T-Tref)-
      T*(prop.Sf+prop.Cp*log(T/Tref)-R*log(P/Pref)));
    }

double G() /* Gibbs free energy of all products combined
    */
    {
    int i;
    double G,yS;
    for(yS=i=0;i<nP;i++)
      yS+=yP[i];
    for(G=i=0;i<nP;i++)
      G+=yP[i]*g(pP[i],T2,yP[i]*P2/Pref/yS);
    return(G);
    }
```

The main program consists of several sections, each is described below. The matrices A and B are in bold face.

```
int main(int argc,char**argv,char**envp)
    {
    int i,iter,j,k,n=nP+nE;
    double damp=2.;

/* copy properties into arrays for convenience loops */
/* (these don't change with iterations) */
```

```
  pR[0]=H2;
  pR[1]=O2;
  pP[0]=H;
  pP[1]=H2;
  pP[2]=H2O;
  pP[3]=O;
  pP[4]=O2;
  pP[5]=OH;

/* hydrogen abundance constraints */
/* (these don't change with iterations) */

  C[n*0+0]=1.; /* H */
  C[n*0+1]=2.; /* H2 */
  C[n*0+2]=2.; /* H2O */
  C[n*0+3]=0.; /* O */
  C[n*0+4]=0.; /* O2 */
  C[n*0+5]=1.; /* OH */
  D[0]=2.;     /* 2H2 */

/* oxygen abundance constraints */
/* (these don't change with iterations) */

  C[n*1+0]=0.; /* H */
  C[n*1+1]=0.; /* H2 */
  C[n*1+2]=1.; /* H2O */
  C[n*1+3]=1.; /* O */
  C[n*1+4]=2.; /* O2 */
  C[n*1+5]=1.; /* OH */
  D[1]=2.;     /* O2 */

/* maximum possible product moles */

  yX[0]=2.; /* H */
  yX[1]=1.; /* H2 */
  yX[2]=1.; /* H2O */
  yX[3]=2.; /* O */
  yX[4]=1.; /* O2 */
  yX[5]=2.; /* OH */

/* initial estimate of final temperature */
/* (this will be corrected after each iteration) */

  T2=3400.;

/* initial estimate of product moles */
/* (necessary to initialize iterations) */

  for(i=0;i<nP;i++)
```

```c
    yP[i]=yX[i]/4.;

/* fill in free energy minimization terms */
/* (these change with each iteration) */

  for(iter=0;iter<32;iter++)
    {
    printf("iter=%i, ",iter);
    memset(A,0,sizeof(A)); /* zero out matix A (zero B
    not necessary) */
    for(yS=i=0;i<nP;i++)
      yS+=yP[i];
    for(i=0;i<nP;i++)
      {
      for(j=0;j<nP;j++)
        {
        if(j==i)
          A[n*i+i]=R*T2*(yS-yP[i])/yP[i]/yS;
        else
          A[n*i+j]=-R*T2/yS;
        }
      B[i]=-g(pP[i],T2,yP[i]*P2/Pref/yS);
      }
/* add C, C-transpose, and D */
/* (these don't change, but the matrices A and B are
    destroyed) */

    for(i=0;i<nE;i++)
      {
      k=nP+i;
      for(j=0;j<nP;j++)
        A[n*k+j]=A[n*j+k]=C[n*i+j];
      B[k]=D[i];
      }

/* solve simultaneous linear equations */

    i=GaussElimination(A,B,n);
    printf("error=%i\n",i);
    for(i=0;i<nP;i++)
      {

    yP[i]=fmax(yP[i]/2.,fmin((yP[i]+yX[i])/2.,(damp*yP[i]
    +B[i])/(damp+1.)));
      printf("%-3s=%.12lf\n",Name[i],yP[i]);
      }
    printf("T2=%.0lf, hR=%.0lf,
    hP=%.0lf\n",T2,hR(),hP());
    T2+=(hR+Q-hP())/R/10.;
```

23

```
        }

    return(0);
    }
```

The iterations converge quickly to the following:

```
    iter=0, error=0
    H   =0.268226427478
    H2  =0.243227867135
    H2O=0.358875576660
    O   =0.347908215145
    O2  =0.299491428298
    OH  =0.527566684932
    T2=3400, hR=0, hP=245302
    ...
    iter=15, error=0
    H   =0.039213932879
    H2  =0.086224843513
    H2O=0.738282913751
    O   =0.091056444828
    O2  =0.430673664379
    OH  =0.312136698669
    T2=3152, hR=0, hP=4
    ...
    iter=31, error=0
    H   =0.038921192555
    H2  =0.086396906203
    H2O=0.737554526720
    O   =0.090915137911
    O2  =0.429179067384
    OH  =0.313176499034
    T2=3152, hR=0, hP=0
```

With a few modifications (adding the carbon compounds and changing the hardwired molar quantities), we can solve the methane combustion reaction. The details are in CH4comb.c. The output is:

```
    iter=0, error=0
    C   =0.125000000000
    CH4=0.625000000000
    CO  =0.625000000000
    CO2=0.625073645710
    H   =0.636958447272
    H2  =0.673440746861
    H2O=0.500000000000
    O   =0.250000000000
    O2  =0.754127966024
    T2=3400, hR=-74902, hP=306386
    ...
    iter=15, error=0
    C   =0.000003814697
```

24

```
CH4=0.808614537991
CO =0.196881247008
CO2=0.138903141297
H  =0.320438083603
H2 =1.322077585413
H2O=0.161327217799
O  =0.367560372428
O2 =0.578648752369
T2=3227, hR=-74902, hP=-73520
...
iter=31, error=0
C  =0.000000000242
CH4=0.803735779391
CO =0.196272667097
CO2=0.137654705608
H  =0.317944169445
H2 =1.324319453139
H2O=0.160150232117
O  =0.370713268026
O2 =0.577821982036
T2=3221, hR=-74902, hP=-74908
```

While this explicit coding works well enough, it is far too troublesome to be practical. Clearly, a more flexible program is needed. At the very least, this would include a property database and symbolic interpretation of the reactions. Matrices A and B are simple enough to build once the properties are loaded, but building C and D is more involved. The maximum possible product moles vector, $y_X$, is very important. This is used at each iteration to contain overshoot. The damped and clamped iteration is:

```
yP[i]=fmax(yP[i]/2.,fmin((yP[i]+yX[i])/2.,
      (damp*yP[i]+B[i])/(damp+1.)));
```

is essential to obtain convergence, as the Gauss elimination linear solver often returns wild and extraneous solution vectors, especially in the early iterations.

# Chapter 5. Real Reactants

The easiest assumption to relax is that of a *perfect* gas having constant specific heats. The simplest expression for specific heat would be:

$$C_P = C_1 + C_2 T + C_3 T^2 \qquad (5.1)$$

The enthalpy (still ideas gas) becomes:

$$h = h_F + C_1(T - T_{REF}) + \frac{C_2}{2}(T^2 - T_{REF}^2) + \frac{C_3}{2}(T^3 - T_{REF}^3) \qquad (5.2)$$

The entropy (still ideas gas) becomes:

$$s = s_F + C_1 \ln\left(\frac{T}{T_{REF}}\right) + C_2(T - T_{REF}) + \frac{C_3}{2}(T^2 - T_{REF}^2) \qquad (5.3)$$

These modifications are easily made to the database. Coefficients $C_2$ and $C_3$ can be set to zero for any substance that has constant specific heat (e.g., He, Ne, Ar, Kr, Xe) or for which the variation is not available. Matrices A, C, and D remain unchanged and B is directly calculated from g (or a), which is h-Ts (or u-Ts).

## Residual Enthalpy

Enthalpy is not simply a constant plus the integral of $C_P$ with respect to T. If this were so, there would be no heat of vaporization, as such occurs at constant temperature. We first separate the impact on enthalpy of temperature from the impact of pressure (or density). The result is called *residual* enthalpy. Formulas for residual enthalpy are derived from Maxwell's relationships.[10] The total differential of enthalpy with respect to temperature and pressure can be written:

$$dh = C_P \, dT + \left[V - T\left(\frac{\partial V}{\partial T}\right)_P\right] dP \qquad (5.4)$$

where V is the specific volume. This formula is useless because there is no practical relationship for $\partial V/\partial T$. Fortunately, it can be transformed into a more useful expression. First, we recall from calculus that:

$$d(PV) = PdV + VdP \qquad (5.5)$$

We want to eliminate the VdP term and so replace it with d(PV)-PdV.

$$dh = C_P \, dT - d(PV) + \left[P - T\left(\frac{\partial P}{\partial T}\right)_V\right] dV \qquad (5.6)$$

---

[10] Maxwell's relationships are foundational concepts in thermodynamics and will not be discussed in detail here. The reader is directed other texts for more information.

We can further improve this relationship, dividing by the gas constant, R, and the critical temperature, $T_C$, to non-dimensionalize it. The pressure and specific volume can likewise be non-dimensionalized by the respective critical values. Incorporating the definition of compressibility, $Z=PV/RT$, we finally arrive at something useful:

$$\frac{h_0 - h}{RT_C} = T_R(1-Z) - Z_C \int_{\infty}^{V_R}\left(P_R - T_R\frac{\partial P_R}{\partial T_R}\right)dV_R \qquad (5.7)$$

where $h_0$ is the temperature only dependent portion (Equation 5.2), h is the total enthalpy ($h_0$+residual), $T_R=T/T_C$, $P_R=P/P_C$, $V_R=V/V_C$, and $Z_C=P_CV_C/RT_C$. Expressions for $P(T,V)$ are readily available. These are called Equations of State (EoS).[11] The subscript 0 (viz., $h_0$) indicates the zero-density (or zero-pressure) contribution, which is to say, the portion that depends only on temperature. Residual enthalpy for a van der Waals fluid is shown in the following figure:

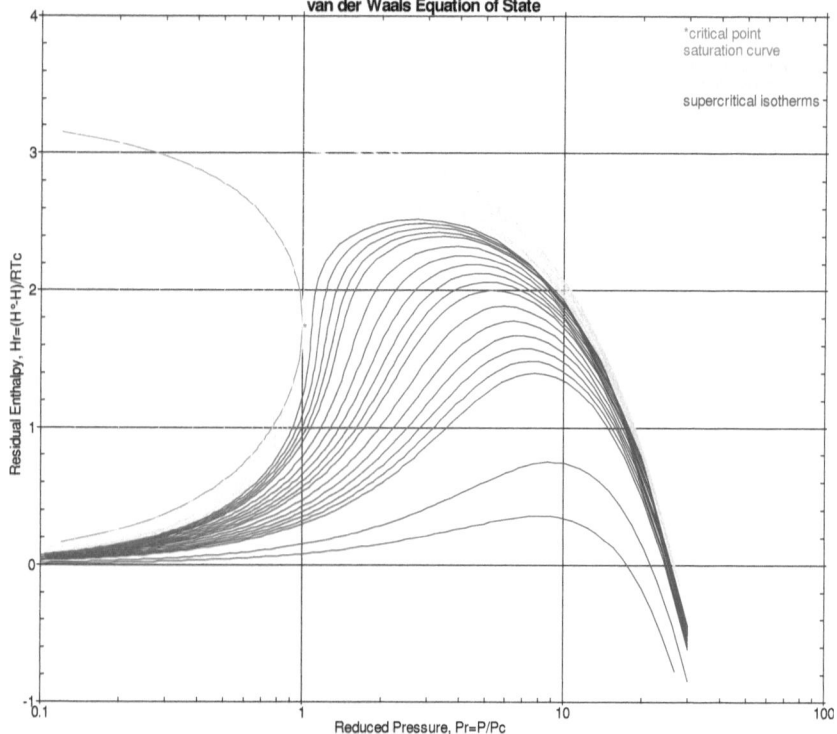

**Figure 7. van der Waals Residual Enthalpy**

---

[11] Equations of state are foundational concepts in thermodynamics and will only be discussed here superficially. The reader is directed other texts for more information.

Residual Entropy

Entropy is also not a simple constant plus the integral of Cp/T with respect to T. We begin this derivation with another inconvenient relationship:

$$ds = C_P \frac{dT}{T} - \left( \frac{\partial V}{\partial T} \right)_P dP \tag{5.8}$$

Through the same transformations, we arrive at:

$$\frac{s_0 - s}{R} = -\ln Z + \int_{\infty}^{V_R} \left( \frac{1}{V_R} - \frac{\partial P_R}{\partial V_R} \right) dV_R \tag{5.9}$$

Residual entropy for a van der Waals fluid is shown in the following figure:

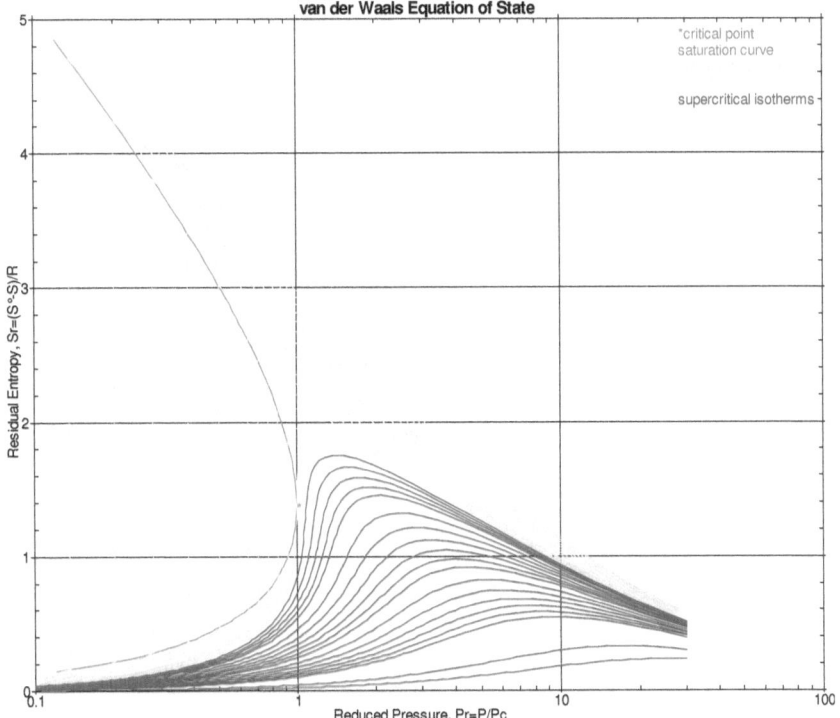

**Figure 8. van der Waals Residual Entropy**

Fugacity

Fugacity (given the symbol, $F$) is a sort of pseudo-pressure. It is often defined as the pressure that would result in an ideal gas having the same Gibbs free energy $(g=h-Ts)$ as the real gas (at the same temperature). We don't ever use fugacity in this way, but it does provide a conceptual framework. Most often, we

29

consider the fugacity coefficient (given the symbol, $\varphi$), which is the ratio of fugacity to pressure, making it dimensionless. There are several derivations and forms used for the fugacity. These derivations are beyond the scope of this work. For our purposes here, we will use the following:

$$\ln \varphi = \int_0^P (Z-1)\frac{dP}{P} \tag{5.10}$$

Equation 5.10 is rarely practical to evaluate. Integration by parts yields the following much more useful formula:

$$\ln \varphi = Z - 1 - \ln Z - \int_\infty^V \left(\frac{P}{RT} - \frac{1}{V}\right) dV \tag{5.11}$$

For an ideal gas $Z=1$ and $\ln\varphi=0$ so that $\varphi=1$ and $F=P$. In general, the less ideal a fluid behaves, the more $Z$ and $\varphi$ depart from unity. The following figure shows the fugacity coefficient for a van der Waals fluid:

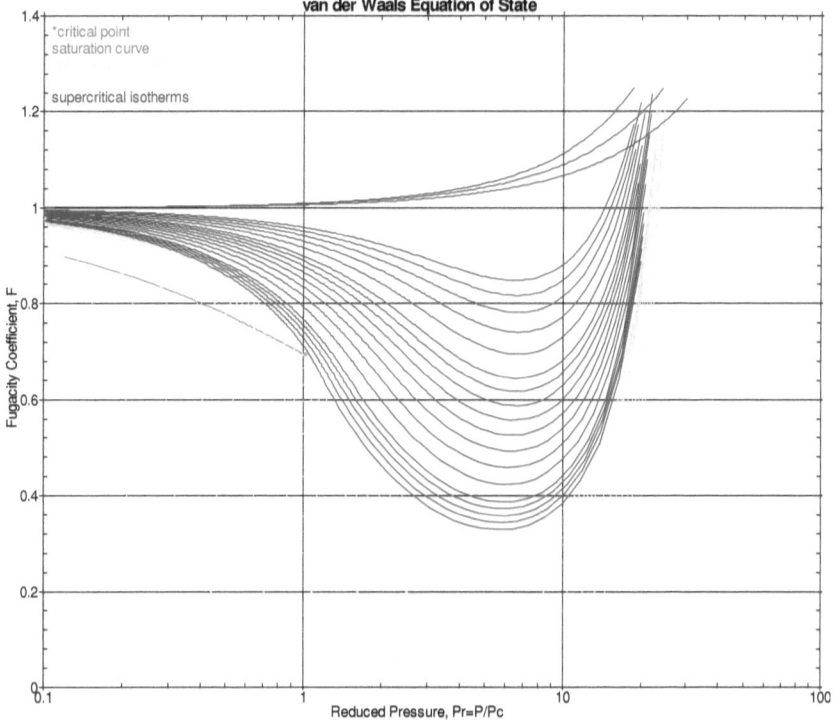

**Figure 9. van der Waals Fugacity Coefficient**

Notice that the red saturation curve has collapsed to a single line (i.e., there is no vapor dome). The fugacity of the saturated liquid and vapor is the same, which ties back into the Gibbs free energy and Maxwell's criterion. Notice also that the sub-critical isotherms (green curves) have the same value, but are not continuous in slope, as they cross the saturation curve.

Combining Equations 5.7, 5.9, and 5.11, we arrive at:

$$\frac{s_0 - s}{R} = \frac{h_0 - h}{RT_C} + \ln F \tag{5.12}$$

### Correcting Matrices A and B

The terms of matrices A and B are defined by Equations 4.10 and 4.11, respectively. The terms of B arise from partial derivatives of $G = \Sigma y_I g_I$. Each of these has two parts: $g_I$ and $y_I(\partial g_I / \partial y_I)$. The first is readily calculated using the formulas derived in this chapter. The second involves Dalton's Partial Pressure Law, which requires two corrections: one direct and one indirect. First, mole fraction has a direct (proportional) impact on partial pressure and so is trivial to calculate. The second arises from the fact that Dalton's Law is only an approximation and assumes the products are gases and behave like hard spheres, which isn't true. We will only correct for the first here and will defer the second correction until the next chapter.

The terms of A arise from the second partial derivatives of G. It is essential to recall at this point that the [A][y]=[B] representation of the free energy minimization problem is actually Newton's Method or the Method of Steepest Descent. To arrive at the correct result, we must minimize the gradient, or $B^TB$ (B transpose times B is the sum of the squares of the terms of B, which are the gradient components). **We don't *have* to solve [A][y]=[B] *exactly*.** The matrix [A] just provides the direction to search for the minimum of [B]. We won't *ever* solve [A][y]=[B] *exactly*. We already forfeited that goal by adding $[C^T][\Lambda]$ to conserve the elements. If we correct [A] at all for these factors (variable specific heat, residual enthalpy, residual entropy, and fugacity), we need only be concerned that the effort is sufficient to facilitate convergence to the solution (i.e., minimized $B^TB$).

The diagonal terms of A are much larger than the off-diagonal terms. This can easily be verified by putting some print statements in H2comb.c or CH4comb.c and recompiling. The diagonal terms scale the step along the descent direction toward the minimum free energy, while the off-diagonal terms fine-tune the path, accounting for the local terrain in our conceptual model.[12] If we simply replace the partial pressures (mole fractions times total pressure) with the fugacities (fugacity coefficients times total pressure) this is adequate for most situations. The off-diagonal terms require only a trivial code adjustment

---

[12] This is just how the Method of Steepest Descent works and is not specific to this application of it.

$(f_I/\Sigma f_I$ vs. $y_I/\Sigma y_I)$, while the diagonal terms require a slightly more complicated one to implement Equation 4.10a.

With these modifications, we are now ready to handle real reactants and can investigate the impact on products and outcomes. Of course, we haven't explained how Equations 5.7, 5.9, and 5.11 can be evaluated. Rather than stop at this point to explain, we will leave that to the next chapter. We solve the reaction on page 10, only account for real reactant properties to obtain:

```
T2=3492.4 K, P2=0.101325, Q=0 kJ
substance        y          x       h/RTo   s/R    g/RTo
---------------------------------------------------------
Hdia          2.0000000 0.6666667   2.8   15.45   -25.0
Odia          1.0000000 0.3333333   2.9   25.23   -42.5
---------------------------------------------------------
reactants     3.0000000 1.0000000   8.5   56.12   -92.5
=========================================================
H             0.1347250 0.0567699 114.6   20.12  -120.9
Hdia          0.3375195 0.1422228  43.2   24.34  -241.7
O             0.0712366 0.0300174 128.7   26.67  -183.5
Odia          0.0990691 0.0417454  47.5   35.41  -367.0
Water         1.4596108 0.6150455 -35.3   33.31  -425.2
Hydroxl       0.2710144 0.1141990  59.4   31.07  -304.4
---------------------------------------------------------
products      2.3731754 1.0000000   8.5   73.37  -850.5
---------------------------------------------------------
difference   -0.6268246 0.0000000  -0.0   17.25  -757.9
```

The differences are not zero, but are trivial. This is to be expected. After all, this reaction occurs at a very high temperature and only one atmosphere pressure. Of course, the products are behaving essentially like ideal gases. But this is not the way we often burn fuels. Gasoline and diesel reciprocating engines and gas turbines typically achieve combustion chamber pressures of 30 to 50 atmospheres, with an average of 40 being common. There is often significant heat transfer out of the combustion chamber (Q<0), whether intentional or unavoidable, resulting in lower product temperatures. Let us consider again the same reaction, only at 40 atmospheres and 1000 degrees less. The following table shows the differences in mole and mass fractions:

| | ideal | | | | real | | | | difference |
|---|---|---|---|---|---|---|---|---|---|
| reactants | Y | X | Z | F | Y | X | Z | F | ΔY |
| Hdia | 2.0000000 | 0.6666667 | 1 | 1 | 2.0000000 | 0.6666667 | 1.04 | 1.05 | 0.0% |
| Odia | 1.0000000 | 0.3333333 | 1 | 1 | 1.0000000 | 0.3333333 | 1.02 | 1.06 | 0.0% |
| products | Y | X | Z | F | Y | X | Z | F | ΔY |
| H | 0.0001502 | 0.0000748 | 1 | 1 | 0.0003219 | 0.0001600 | 1.00 | 1.09 | 114.3% |
| Hdia | 0.0106110 | 0.0052873 | 1 | 1 | 0.0173209 | 0.0086114 | 1.00 | 1.09 | 63.2% |
| O | 0.0000590 | 0.0000294 | 1 | 1 | 0.0001238 | 0.0000615 | 1.00 | 1.08 | 109.8% |
| Odia | 0.0038618 | 0.0019242 | 1 | 1 | 0.0062475 | 0.0031060 | 1.00 | 1.08 | 61.8% |

| | Y | X | Z | F | Y | X | Z | F | ΔY |
|---|---|---|---|---|---|---|---|---|---|
| Water | 1.9864102 | 0.9897906 | 1 | 1 | 1.9776549 | 0.9832254 | 1.03 | 1.02 | -0.4% |
| Hydroxl | 0.0058073 | 0.0028936 | 1 | 1 | 0.0097264 | 0.0048356 | 1.00 | 1.09 | 67.5% |

The compressibilities (Z) and fugacity coefficients (F) are also shown in this table. The differences in Z and F are small, but look at the differences in moles in the right column. If this were a combustion turbine and you were concerned with carbon monoxide and NOx emissions, accounting for real behavior becomes a big deal. We will next consider combustion with carbon and air. Even in this reaction, the quantities of H and OH would be problematic, as these form acids and bases that eat up your machinery. In this case the two OH:H ratios are 39:1 and 30:1, respectively. Note that there is hardly any difference in $H_2O$ (-0.4%) and that the smaller products are impacted more significantly. We next compare combustion of methane:

| | ideal | | | | real | | | | difference |
|---|---|---|---|---|---|---|---|---|---|
| reactants | Y | X | Z | F | Y | X | Z | F | ΔY |
| Methane | 1.000000 | 0.333333 | 1 | 1 | 1.000000 | 0.333333 | 1.0279 | 1.0808 | 0.0% |
| Odia | 2.000000 | 0.666667 | 1 | 1 | 2.000000 | 0.666667 | 1.0416 | 1.0605 | 0.0% |
| products | Y | X | Z | F | Y | X | Z | F | ΔY |
| H | 0.000145 | 0.000048 | 1 | 1 | 0.000297 | 0.000098 | 1.0000 | 1.0787 | 105.1% |
| Hdia | 0.006559 | 0.002172 | 1 | 1 | 0.010422 | 0.003439 | 1.0002 | 1.0787 | 58.9% |
| O | 0.000144 | 0.000048 | 1 | 1 | 0.000283 | 0.000093 | 1.0000 | 1.0718 | 96.7% |
| Odia | 0.015246 | 0.005048 | 1 | 1 | 0.022640 | 0.007471 | 1.0005 | 1.0718 | 48.5% |
| Water | 1.988833 | 0.658554 | 1 | 1 | 1.982283 | 0.654172 | 1.0166 | 1.0254 | -0.3% |
| Hydroxl | 0.009072 | 0.003004 | 1 | 1 | 0.014292 | 0.004717 | 1.0003 | 1.0801 | 57.5% |
| C | 0.000000 | 0.000000 | 1 | 1 | 0.000000 | 0.000000 | 1.0000 | 1.0565 | 0.0% |
| Cmonox | 0.028540 | 0.009450 | 1 | 1 | 0.042138 | 0.013906 | 1.0011 | 1.0880 | 47.6% |
| Cdiox | 0.971460 | 0.321676 | 1 | 1 | 0.957862 | 0.316103 | 1.0233 | 1.0724 | -1.4% |

Again, there is little difference in $H_2O$ (-0.3%) and $CO_2$ (-1.4%). The OH:H ratios are now 63 and 48, respectively. In all of these cases, the fugacity coefficients are greater than unity. This means that the products are more active (and sensitive) than ideal behavior. This is not always the case, though for combustion reactions and high temperatures it is a safe assumption.

# Chapter 6. Real Fluids

In developing relationships for residual enthalpy, entropy, and fugacity in Chapter 5, we were careful to formulate these in terms of P(T,V). All practical Equations of State (EoS) are cast in this form. This facilitates evaluation of Equations 5.7, 5.9, and 5.10. The most common (and earliest significant) EoS was proposed by van der Waals:[13]

$$P = \frac{RT}{V-b} - \frac{a}{V^2} \tag{6.1}$$

The first term, RT/(V-b), accounts for finite molecular size and is a repulsive term. The constant b is representative of the hard sphere volume. The second term, $a/V^2$, accounts for the attractive force (often called the van der Waals force), that provides cohesion and fluid continuity. This simple equation is surprisingly accurate, especially considering the paucity of data available at the time of its introduction.

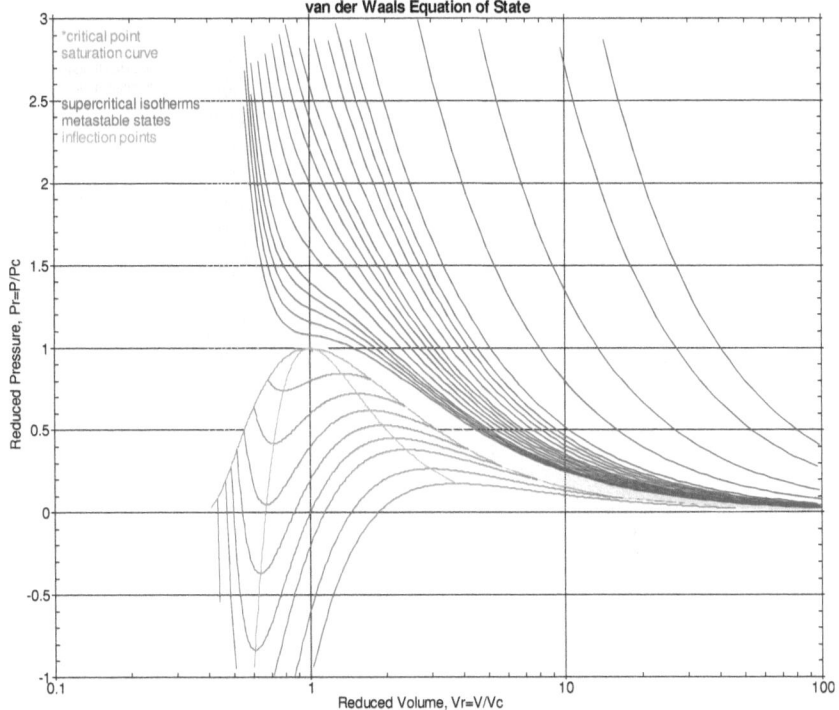

**Figure 10. van der Waals Vapor Dome**

---

[13] Johannes Diderik van der Waals (1837–1923) Dutch theoretical physicist and thermodynamicist.

It is convenient to express equations of state in reduced (dimensionless) parameters, including $P_R=P/P_C$, $V_R=V/V_C$, and $T_R=T/T_C$. The preceding graph is drawn with these dimensionless parameters. Equation 6.1 can be expressed in dimensionless form:

$$Z_C P_R = \frac{T_R}{V_R - B} - \frac{A}{V_R^2} \tag{6.2}$$

The residual enthalpy, entropy, and fugacity were illustrated in the preceding chapter. We can substitute 6.2 into 5.7 and 5.10 to obtain:

$$\frac{h_0 - h}{RT_C} = T_R(1 - Z) + \frac{A}{V_R} \tag{6.3}$$

$$\ln \phi = Z - 1 - \ln Z + \ln\left(\frac{V_R}{V_R - B}\right) - \frac{A}{V_R T_R} \tag{6.4}$$

The residual entropy is calculated from Equation 5.11 using the results of 6.3 and 6.4, that is $(s_0-s)/R=(h_0-h)/RT_C+\ln(\varphi)$. These relationships are used to adjust matrices A and B in order to obtain results closer to actual fluid behavior. The compressibility ($Z=PV/RT$) of a van der Waals fluid is shown below:

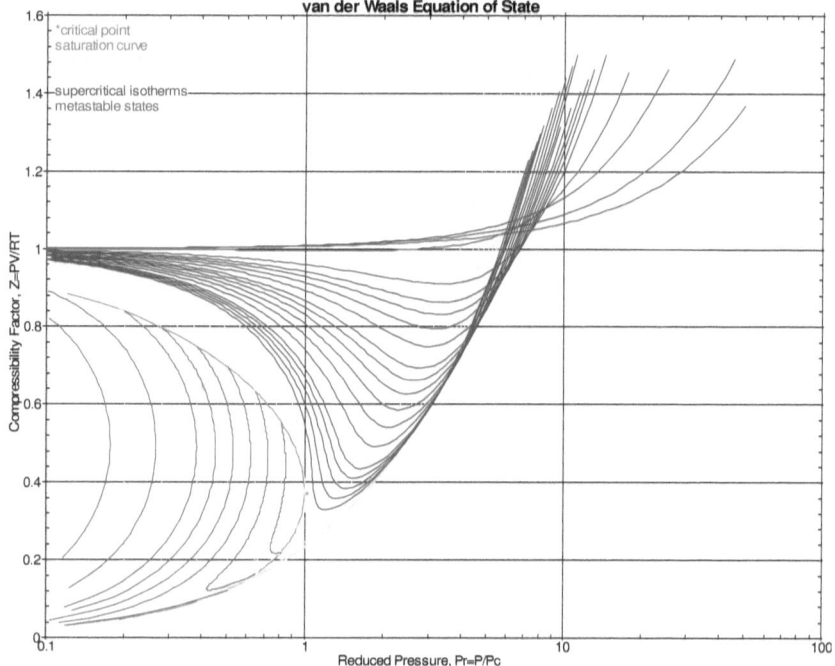

**Figure 11. van der Waals Compressibility Factor**

Consider the combustion of natural gas with air at 40 atm and 2500°K.

| T2=2500 K, P=40 atm, Q/RTo=-2178 | | | | | | | | |
|---|---|---|---|---|---|---|---|---|
| | ideal | | | | real | | | diff. |
| reactants | Y | X | Z | F | Y | X | Z | F | ΔY |
| Methane | 0.757813 | 0.021050 | 1 | 1 | 0.757813 | 0.021050 | 1.0017 | 1.0814 | 0.0% |
| Ethane | 0.125000 | 0.003472 | 1 | 1 | 0.125000 | 0.003472 | 1.0004 | 1.1206 | 0.0% |
| Propane | 0.062500 | 0.001736 | 1 | 1 | 0.062500 | 0.001736 | 1.0003 | 1.1703 | 0.0% |
| Butane | 0.031250 | 0.000868 | 1 | 1 | 0.031250 | 0.000868 | 1.0002 | 1.2228 | 0.0% |
| Pentane | 0.015625 | 0.000434 | 1 | 1 | 0.015625 | 0.000434 | 1.0001 | 1.2864 | 0.0% |
| Hexane | 0.007813 | 0.000217 | 1 | 1 | 0.007813 | 0.000217 | 1.0001 | 1.3645 | 0.0% |
| Air | 35.000000 | 0.972222 | 1 | 1 | 35.000000 | 0.972222 | 1.0766 | 1.0755 | 0.0% |
| products | Y | X | Z | F | Y | X | Z | F | ΔY |
| Ar | 0.164500 | 0.008740 | 1 | 1 | 0.164500 | 0.008736 | 1.0006 | 1.0626 | 0.0% |
| C | 0.000000 | 0.000000 | 1 | 1 | 0.000000 | 0.000000 | 1.0000 | 1.0417 | 0.0% |
| Cdiox | 1.438568 | 0.076433 | 1 | 1 | 1.430228 | 0.075955 | 1.0055 | 1.0726 | -0.6% |
| Cmonox | 0.013744 | 0.000730 | 1 | 1 | 0.022085 | 0.001173 | 1.0001 | 1.0770 | 60.7% |
| H | 0.000228 | 0.000012 | 1 | 1 | 0.000441 | 0.000023 | 1.0000 | 1.0601 | 93.7% |
| Hdia | 0.002596 | 0.000138 | 1 | 1 | 0.004422 | 0.000235 | 1.0000 | 1.0601 | 70.3% |
| Hydroxl | 0.043864 | 0.002331 | 1 | 1 | 0.058176 | 0.003090 | 1.0002 | 1.0661 | 32.6% |
| O | 0.002760 | 0.000147 | 1 | 1 | 0.004177 | 0.000222 | 1.0000 | 1.0616 | 51.4% |
| Odia | 0.900238 | 0.047831 | 1 | 1 | 0.859296 | 0.045634 | 1.0028 | 1.0616 | -4.5% |
| Ozone | 0.000000 | 0.000000 | 1 | 1 | 0.000001 | 0.000000 | 1.0000 | 1.0604 | 104.4% |
| Water | 2.420670 | 0.128614 | 1 | 1 | 2.411582 | 0.128071 | 1.0029 | 1.0374 | -0.4% |
| N | 0.000000 | 0.000000 | 1 | 1 | 0.000001 | 0.000000 | 1.0000 | 1.0759 | 338.1% |
| Ndia | 13.619920 | 0.723648 | 1 | 1 | 13.578750 | 0.721123 | 1.0567 | 1.0759 | -0.3% |
| Ndiox | 0.001736 | 0.000092 | 1 | 1 | 0.003070 | 0.000163 | 1.0000 | 1.0681 | 76.8% |
| Nitrousoxide | 0.000047 | 0.000002 | 1 | 1 | 0.000133 | 0.000007 | 1.0000 | 1.0636 | 184.4% |
| Nmonox | 0.212328 | 0.011281 | 1 | 1 | 0.293158 | 0.015569 | 1.0008 | 1.0547 | 38.1% |
| Nndiox | 0.000000 | 0.000000 | 1 | 1 | 0.000001 | 0.000000 | 1.0000 | 1.0636 | 254.6% |
| Nnpentaox | 0.000000 | 0.000000 | 1 | 1 | 0.000000 | 0.000000 | 1.0000 | 1.0636 | 0.0% |
| Nntetraox | 0.000000 | 0.000000 | 1 | 1 | 0.000000 | 0.000000 | 1.0000 | 1.0636 | 0.0% |
| Nntriox | 0.000000 | 0.000000 | 1 | 1 | 0.000000 | 0.000000 | 1.0000 | 1.0636 | 347.7% |
| Ntriox | 0.000000 | 0.000000 | 1 | 1 | 0.000000 | 0.000000 | 1.0000 | 1.0636 | 240.1% |
| NOx | 0.214111 | 0.011376 | 1 | 1 | 0.296361 | 0.015739 | 1.0008 | 1.0548 | 38.5% |

The mole-weighted NOx is calculated along the bottom row. Correcting the fugacities predicts an increase of 38.5%. The real and ideal OH:H ratios are 193 and 132, respectively. Note that 60.7% more carbon monoxide is predicted.

## Impact of Temperature on Composition

We will next consider the impact of temperature on composition, that is, the predicted moles of each product.

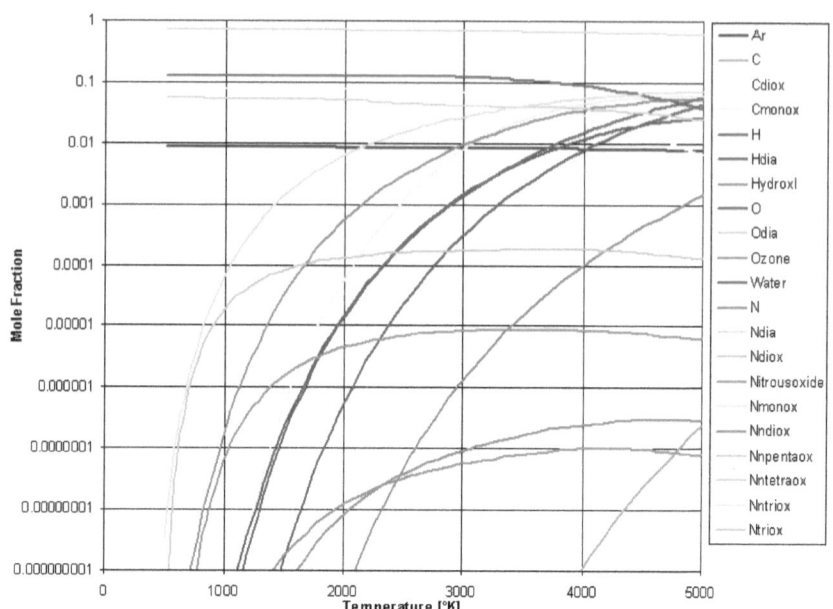

**Figure 12. Impact of Temperature on Combustion Products of Natural Gas**

We can isolate the OH:H ratios:

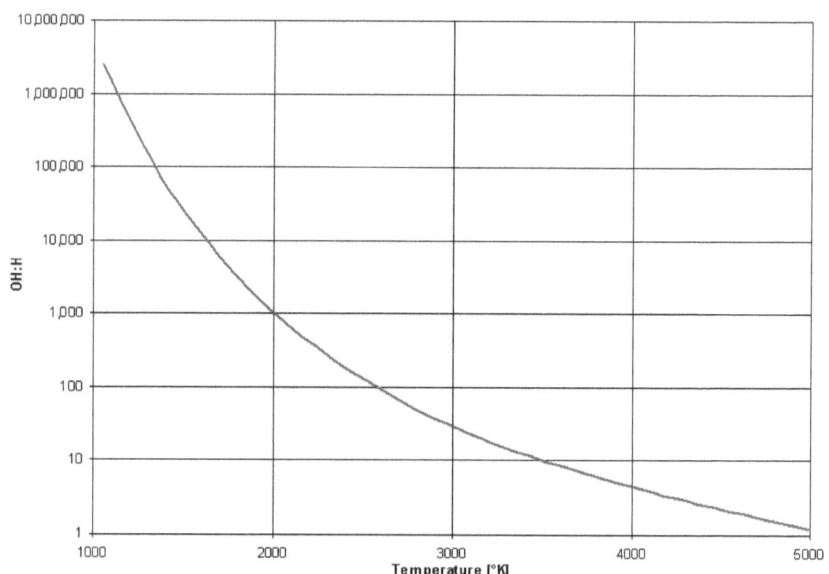

**Figure 13. Impact of Temperature on OH:H Ratio**

We can also isolate the total moles of NOx:

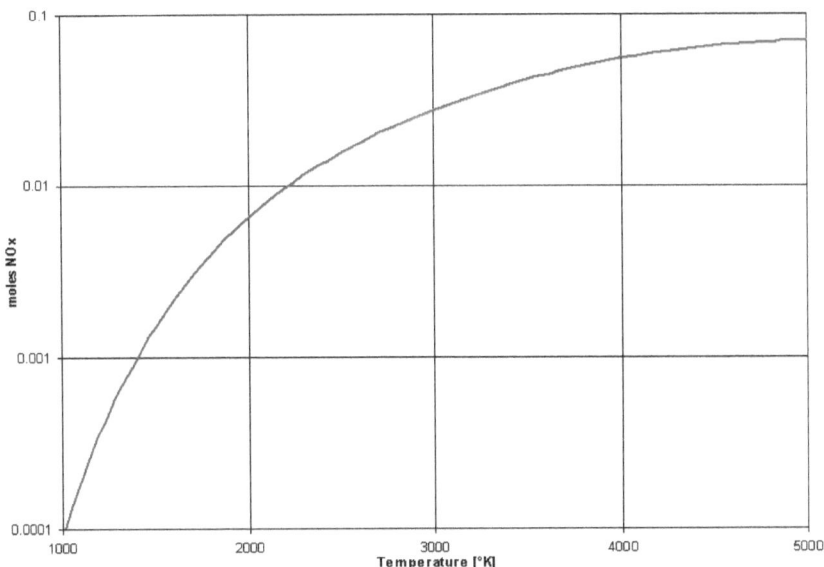

**Figure 14. Impact of Temperature on NOx**

And the moles of carbon monoxide:

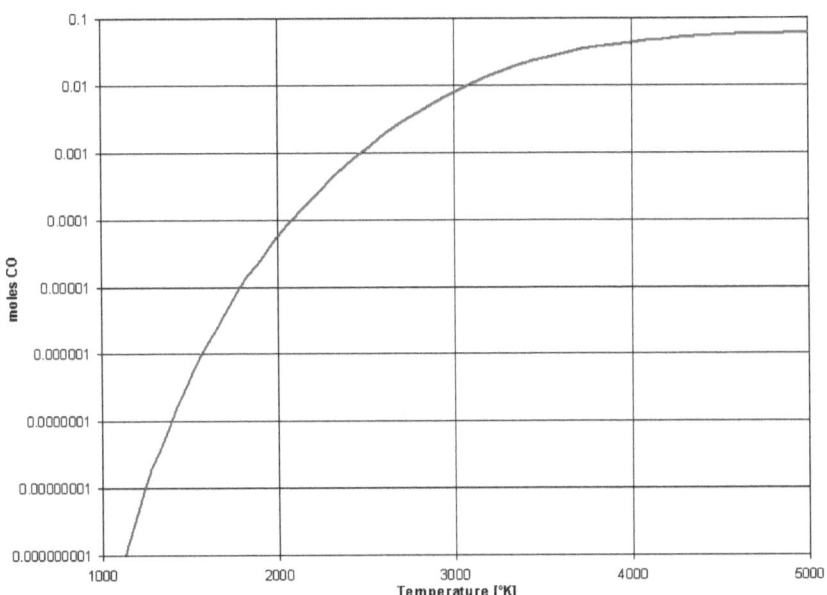

**Figure 15. Impact of Temperature on CO**

## Impact of Pressure on Composition

The convenience of having a program that reads properties from a database, interprets reactions symbolically, calculates a sequence of reactions, creates a graph, and copies the results to the clipboard for pasting into Excel® or Word® should be readily apparent by this point. We can use this program (CREST) to solve the combustion of methane and sweep across a range of pressures to obtain the following graph of mole fractions:

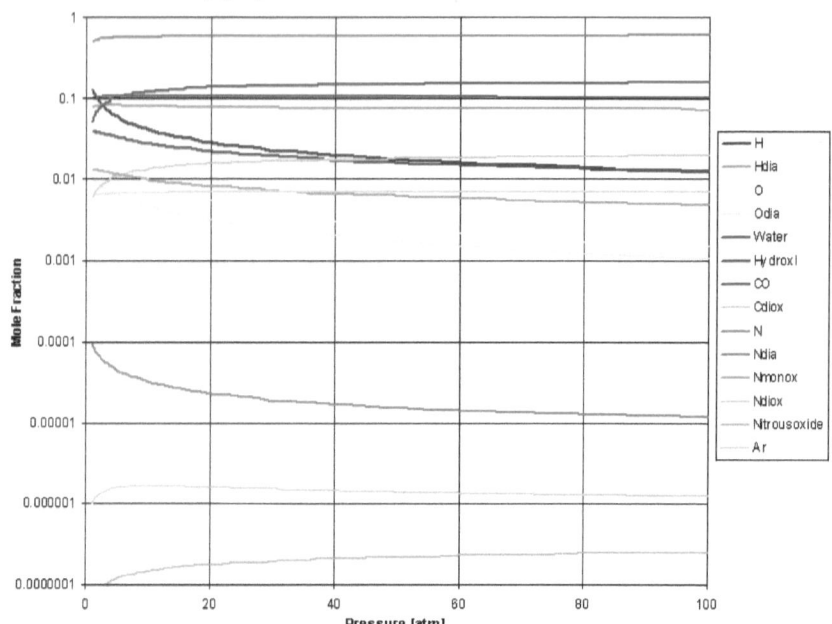

**Figure 16. Impact of Pressure on Combustion of Methane**

While there doesn't appear to be a strong influence of pressure, remember that the Y-axis is logarithmic. Note that some products ($H_2O$, $CO_2$, and $N_2$) increase with pressure, while others (H, $H_2$, O, OH, NO, NOx, and OH:H) decrease. This is typical for combustion. The larger, more stable molecules are preferred over the smaller, particularly the dissociated ones. The thermodynamic reason for this trend arises from the product $y_I g_I$. The Gibbs free energy, $g_I$, contains the term, $-Ts$, and the term, s, contains the term, $-R\ln(P_I)$, which expands to $+RT\ln(P_I)$. Fewer, larger molecules result in smaller values of $\Sigma y_I$ and less contribution from $RT\ln(P_I)$ and lower free energy.

For this reaction, we see only a slight impact of pressure on CO, a slight decrease in NOx, and a slight increase in OH:H. This follows directly from the preceding relationship between these molecules, their molar quantities, and the free energy of each one. The enthalpy and entropy of formation for these

40

products is less important for this particular scenario than the influence of Rln(P). These last three quantities are illustrated separately in this next figure.

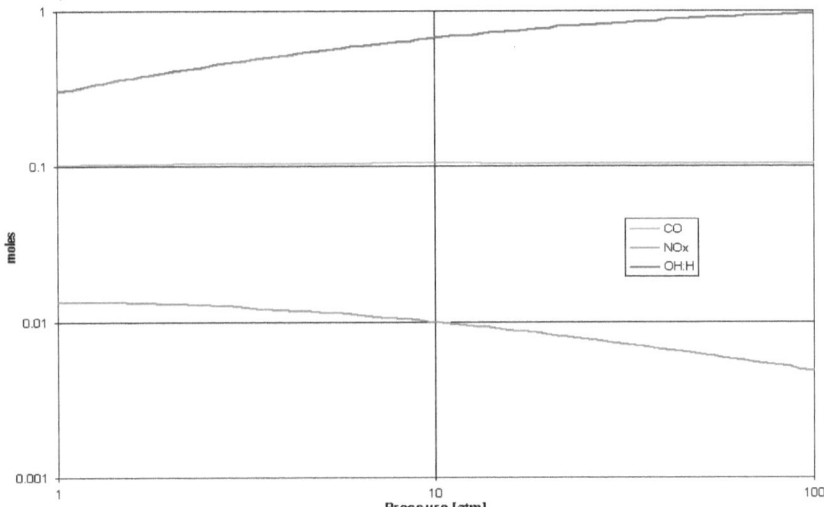

**Figure 17. Impact of Pressure on CO, NOx, and OH:H Ratio**

We see these same trends even in a simpler reaction, previously solved as ideal: combustion of methane with excess oxygen:

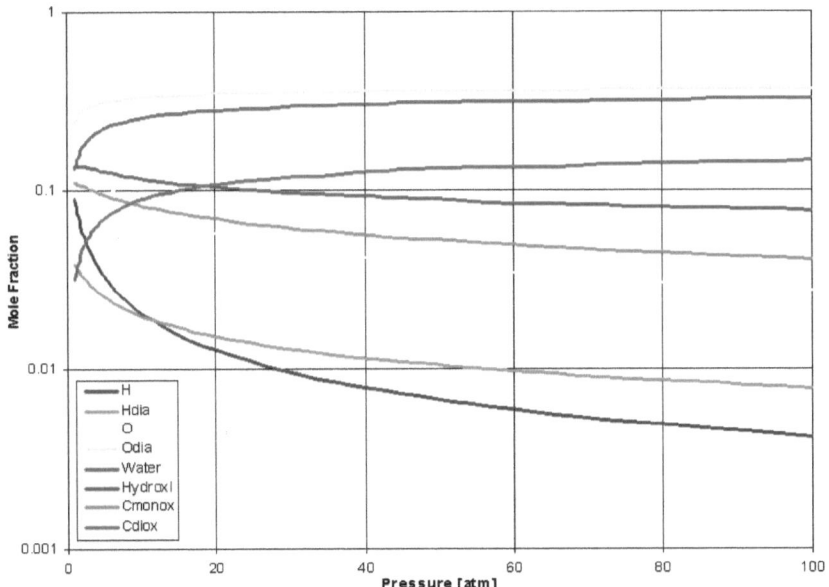

**Figure 18. Impact of Pressure on Simplified Methane Combustion**

41

As evidence that the formulation presented here adequately predicts the compressibility of mixtures, the following comparison to experimental and previous analytical results are provided. Data is for a mixture of nitrogen and ethylene at 50°C. First the previous work:[14]

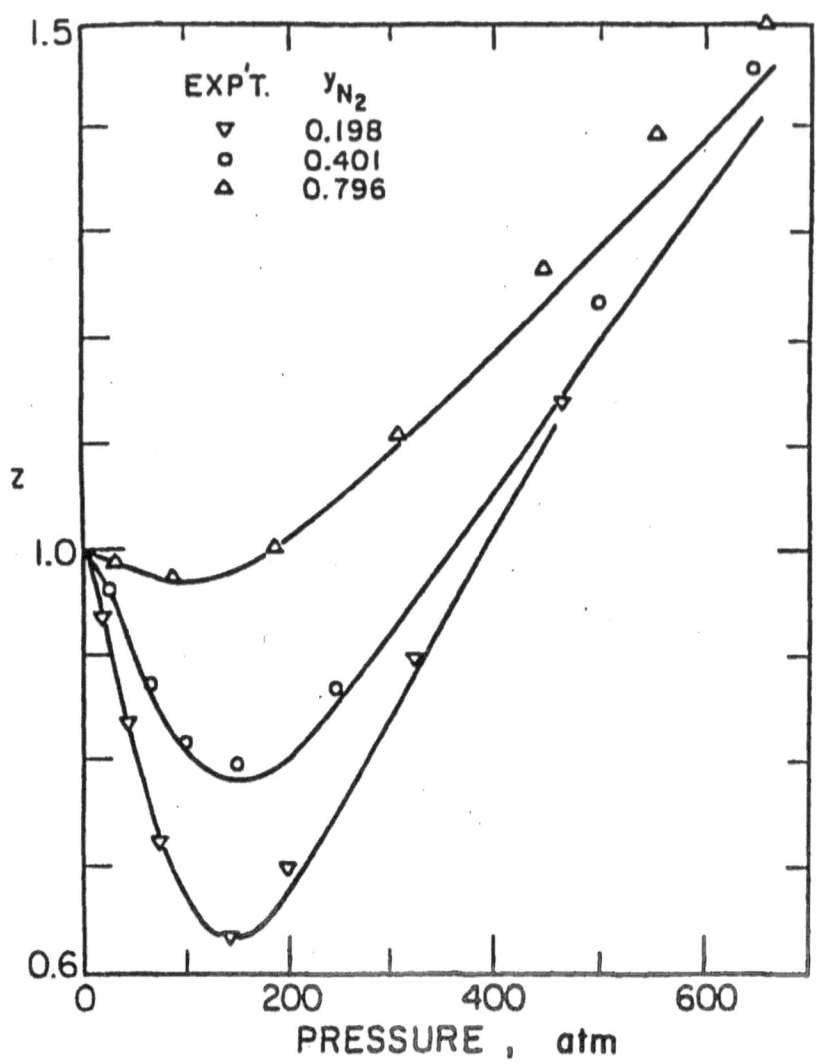

Figure 19. Mixture of Nitrogen and Ethylene (prior)

---

[14] I am sorry to say that I can't remember where this figure came from or I would be most pleased to give credit where it is due.

The same calculations performed with the formulation described herein and built into the CREST code is:

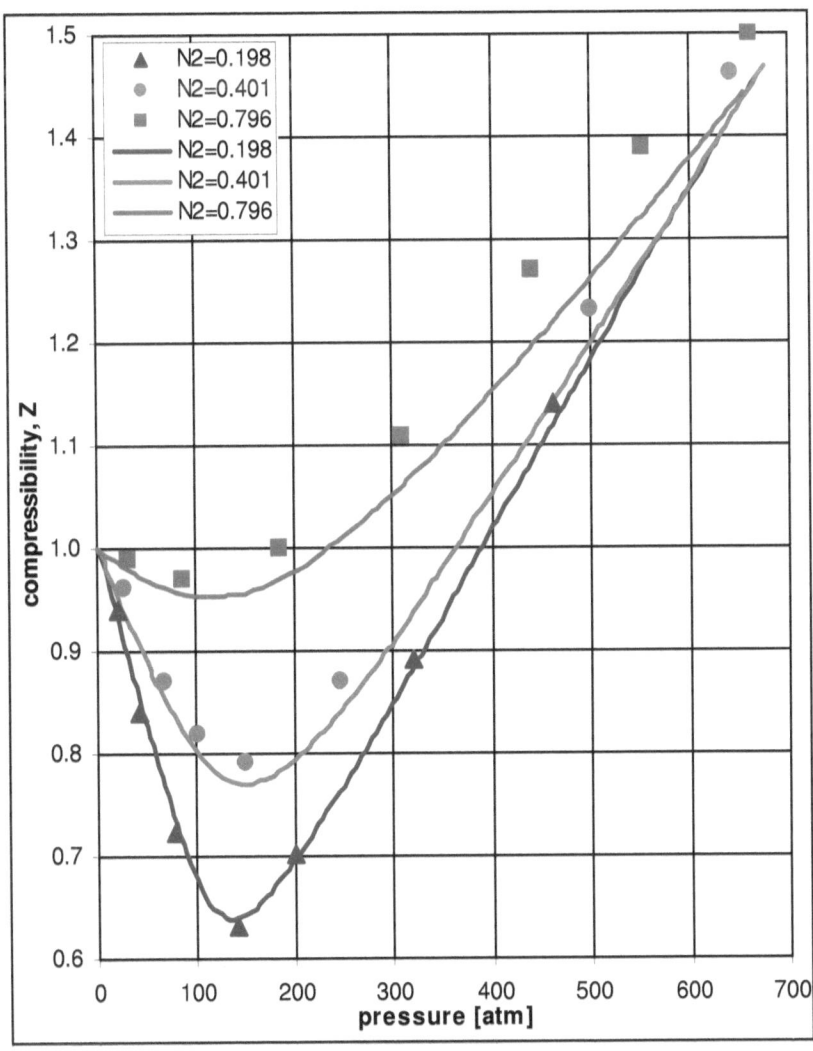

**Figure 20. Mixture of Nitrogen and Ethylene (this work)**

The following comparison of fugacity coefficients for methane in hydrogen sulfide and ethane is provided as evidence that the formulation presented herein accurately represents similar results reported in the literature.[15]

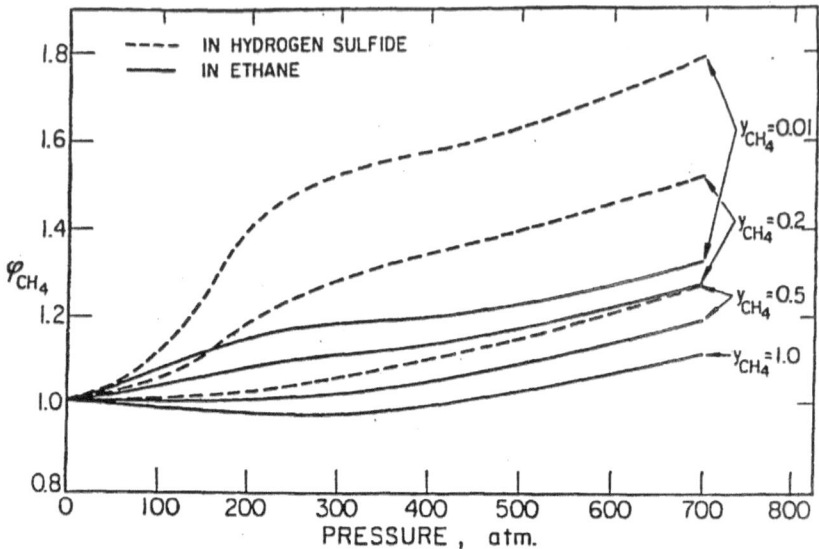

Figure 21. Mixture of Methane, Ethane, and $H_2S$ (prior)

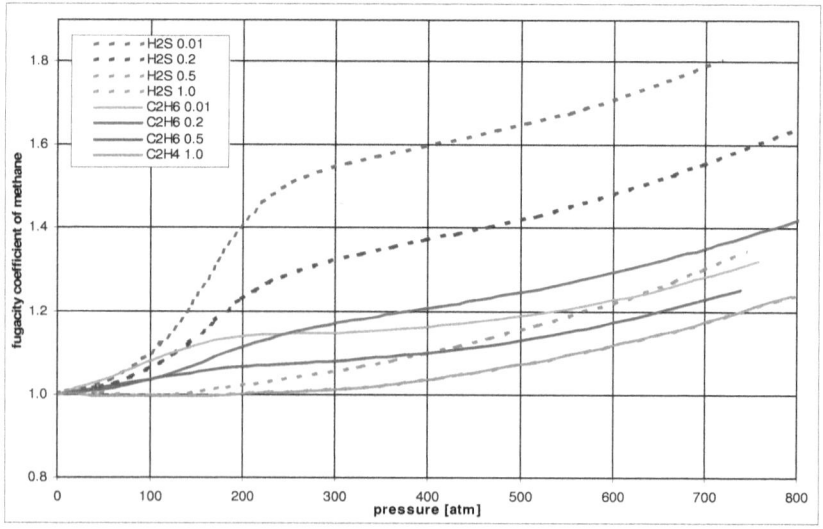

Figure 22. Mixture of Methane, Ethane, and $H_2S$ (this work)

---

[15] Regrettably, I don't recall the source.

# Chapter 7. Real Mixtures

The first successful practical treatment of mixtures was published by Redlich and Kwong in 1949.[16] This work contained a slight modification of the van der Waals equation that constituted a significant improvement:

$$P = \frac{RT}{(V-b)} - \frac{a}{V(V+b)\sqrt{T}}$$

$$a = 0.42748\frac{R^2 T_C^{2.5}}{P_C} \tag{7.1}$$

$$b = 0.08664\frac{RT_C}{P_C}$$

The parameters a and b are somewhat empirical, making $Z_C=1/3$, which is too large, but we're not concerned with that here. The most important contribution of Redlich and Kwong were their algorithms for combining multiple fluids into a single pseudo-fluid. These algorithms are called *mixing rules*. Soave published a modification to the $\sqrt{T}$ term in 1972, but did not change the mixing rules, which are of primary interest here.[17] This later work did serve to further demonstrate the power of this simple EoS to predict the behavior of real fluids, which is why we use it to estimate real fugacity coefficients and compressibilities. The R-K mixing rules are equations for calculating a and b of the pseudo-fluid.

$$b = \sum_I x_I b_I$$

$$a = \sum_I \sum_J x_I x_J a_{I,J} \tag{7.2}$$

$$a_{I,J} = \sqrt{a_I a_J}$$

where $x_I$ are the mass fractions (not mole fractions). The critical temperature and pressure of the pseudo-fluid are simply mass-weighted. Authors differ on how to best calculate the critical volume. Some suggest simple mass weighting, while others mass-weight the cube-roots, which is like averaging the radii of hard spheres. Still other authors recommend mass-weighting the critical compressibilities, $Z_C$, and then calculating the critical volume. This last approach is used here.

---

[16] Redlich, O. and Kwong, J. N. S., "On The Thermodynamics of Solutions," *Chemical Review*, Vol. 44, No. 1, pp. 233–244, 1949.
[17] Soave, G., "Equilibrium Constants from a Modified Redlich-Kwong Equation of State," *Chemical Engineering Science*, Vol. 27, No. 6, pp. 1197–1203, 1972.

Once the properties of the pseudo-fluid have been calculated, the residual enthalpy, entropy, fugacity coefficient, and compressibility are easily calculated and quite similar to the van der Waals values presented in Chapter 5. Suffice it to say, this is not the place to dwell on the efficacy of these mixing rules. Others have undertaken to do this and the reader is directed there for further reading.[18]

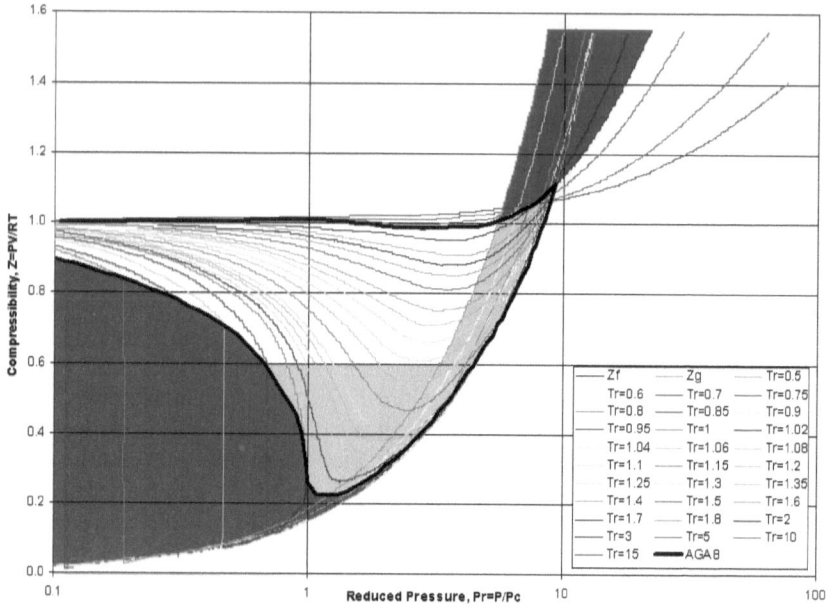

**Figure 23. Compressibility Chart for Natural Gas**

Accurate data for complex mixtures of fluids is quite scarce, owing to the great expense and limited motivation. The most significant contribution in this area comes from the American Gas Association. Accurate quantification of natural gas, which is a mixture of fluids, is a multi-billion dollar concern. Of course, they would invest the money necessary to conduct the research and develop the equations.[19] This valuable work is contained in AGA Report No.

---

[18] Prausnitz, J. M., Lichtenthaler, R. N., and de Azevedo, E. G., *Molecular Thermodynamics of Fluid-Phase Equilibria, 3rd Edition*, Prentice Hall, 1999.

[19] Note of historical interest: Dmitri Ivanovich Mendeleev (1834-1907) Russian chemist and inventor, formulated the Periodic Law, and created an early version of the periodic table of elements. He was financially motivated to accurately quantify the value of coal from various sources. In other words, he developed the science to assure people were getting their money's worth and not being cheated!

8.[20] This material is so valuable that NIST has a web page devoted to it and even provides a spreadsheet:

https://pages.nist.gov/AGA8/

I have programmed the entire AGA8 formulation and have an Excel® Add-In that implements it quite efficiently. This software does, however, belong to my previous employer, McHale Performance, who sells it, so I'm not going to give it away. Still, I have provided a spreadsheet (natural_gas.xls) in the examples\natural gas folder that implements the R-K mixing rules and compares this to AGA8 for your edification. The AGA8 formulation only covers the region inside the heavy black outline in the preceding figure. The blue-shaded region, which are the liquids (compressed and saturated), are not covered. This spreadsheet provides several other calculations, including dew-point of the mixture, something of interest to gas turbine operators.

| | name | Mole Fraction | formula | MW | LHV BTU/lb-mole | HHV BTU/lb-mole | Tc °R | Pc psia | Vc ft3/lbm | Zc | Partial-Pres. psia | Vapor Pres. psia | test cond-sed |
|---|---|---|---|---|---|---|---|---|---|---|---|---|---|
| | **Natural Gas Mole Fractions and Properties** | | | | | | | | | | | | |
| | (some water vapor) | | | | | | | | | | 58 temperature [°F] | | |
| | | | | | | | | | | | 600.000 pressure [psia] | | |
| 1 | Methane | 91.900% | CH4 | 16.043 | 909.4 | 1010.0 | 343.0 | 667.8 | 0.09888 | 0.2878 | 551.400 | N/A | N/A |
| 2 | Nitrogen | 4.220% | N2 | 28.013 | 0.0 | 0.0 | 227.2 | 492.8 | 0.05151 | 0.2916 | 25.320 | N/A | N/A |
| 3 | Carbon Dioxide | 1.050% | CO2 | 44.010 | 0.0 | 0.0 | 548.2 | 1070.6 | 0.03423 | 0.2741 | 6.300 | N/A | N/A |
| 4 | Ethane | 1.810% | C2H6 | 30.070 | 1618.7 | 1769.6 | 549.8 | 707.8 | 0.07891 | 0.2846 | 10.860 | 484.932 | no |
| 5 | n-Propane | 0.610% | C3H8 | 44.097 | 2314.9 | 2516.1 | 665.7 | 616.3 | 0.07382 | 0.2808 | 3.660 | 104.959 | no |
| 6 | Water | 0.040% | H2O | 18.015 | 0.0 | 0.0 | 1165.2 | 3208.1 | 0.04929 | 0.2278 | 0.240 | 0.239 | YES |
| 7 | Hydrogen Sulfide | 0.000% | H2S | 34.082 | 586.8 | 637.1 | 672.4 | 1306.5 | 0.05167 | 0.3188 | 0.000 | N/A | N/A |
| 8 | Hydrogen | 0.000% | H2 | 2.016 | 273.8 | 324.2 | 59.8 | 188.1 | 0.51672 | 0.3053 | 0.000 | N/A | N/A |
| 9 | Carbon Monoxide | 0.000% | CO | 28.010 | 320.5 | 320.5 | 239.3 | 507.5 | 0.05322 | 0.2946 | 0.000 | N/A | N/A |
| 10 | Oxygen | 0.000% | O2 | 31.999 | 0.0 | 0.0 | 278.6 | 736.9 | 0.03823 | 0.3015 | 0.000 | N/A | N/A |
| 11 | i-Butane | 0.120% | C4H10 | 58.123 | 3000.4 | 3251.9 | 734.6 | 529.1 | 0.07248 | 0.2827 | 0.720 | 36.845 | no |
| 12 | n-Butane | 0.160% | C4H10 | 58.123 | 3010.8 | 3262.3 | 765.3 | 554.0 | 0.07026 | 0.2755 | 0.960 | 25.142 | no |
| 13 | i-Pentane | 0.040% | C5H12 | 72.150 | 3699.0 | 4000.9 | 828.7 | 490.4 | 0.06787 | 0.2700 | 0.240 | 9.012 | no |
| 14 | n-Pentane | 0.020% | C5H12 | 72.150 | 3706.9 | 4008.9 | 845.4 | 490.1 | 0.06759 | 0.2634 | 0.120 | 6.640 | no |
| 15 | n-Hexane | 0.030% | C6H14 | 86.177 | 3856.6 | 4205.4 | 913.2 | 430.6 | 0.06875 | 0.2603 | 0.180 | 1.838 | no |
| 16 | n-Heptane | 0.000% | C7H16 | 100.204 | 4465.7 | 4864.3 | 972.3 | 396.8 | 0.06905 | 0.2631 | 0.000 | 0.510 | no |
| 17 | n-Octane | 0.000% | C8H18 | 114.231 | 5074.9 | 5523.4 | 1023.8 | 360.6 | 0.06905 | 0.2589 | 0.000 | 0.146 | no |
| 18 | n-Nonane | 0.000% | C9H20 | 128.258 | 5683.2 | 6181.6 | 1070.2 | 335.1 | 0.06630 | 0.2481 | 0.000 | 0.044 | no |
| 19 | n-Decane | 0.000% | C10H22 | 142.285 | 6294.7 | 6842.8 | 1111.9 | 304.6 | 0.06736 | 0.2447 | 0.000 | 0.013 | no |
| 20 | Helium | 0.000% | He | 4.003 | 0.0 | 0.0 | 9.3 | 33.2 | 0.23115 | 0.3078 | 0.000 | N/A | N/A |
| 21 | Argon | 0.000% | Ar | 39.948 | 0.0 | 0.0 | 271.5 | 706.9 | 0.02989 | 0.2897 | 0.000 | N/A | N/A |
| | Composite | 100.000% | | 17.440 | 15,661 | 17,343 | 347.9 | 649.2 | 0.09486 | 0.2876 | 600.000 | | |

The aforementioned calculations are illustrated in the following VBA® code that can be found in the spreadsheet:

```
Function CompositeCriticalTemperature(xMole As Range) As
    Double
    'this function returns the critical temperature in °R of
        the mixture given the mole fractions
    Dim i As Integer
```

---

[20] Starling, K. E. and Savidge, J. L., "Compressibility Factors of Natural Gas and Other Related Hydrocarbon Gases," American Gas Association Transmission Measurement Committee Report No. 8, 1992 (revised 1994).

```
  CompositeCriticalTemperature = 0#
  For i = 1 To 21
    CompositeCriticalTemperature =
    CompositeCriticalTemperature + xMole(i) *
    CriticalTemperature(i - 1)
  Next i
End Function
Function CompositeCriticalCompressibility(xMole As
    Range) As Double
'this function returns the critical compressibility of
    the mixture given the mole fractions
  Dim i As Integer
  CompositeCriticalCompressibility = 0#
  For i = 1 To 21
    CompositeCriticalCompressibility =
    CompositeCriticalCompressibility + xMole(i) *
    CriticalCompressibility(i - 1)
  Next i
End Function
Function CompositeCriticalVolume(xMole As Range) As
    Double
'this function returns the critical specific volume in
    ft^3/lbm of the mixture given the mole fractions
  Dim i As Integer
  CompositeCriticalVolume = 0#
  For i = 1 To 21
    CompositeCriticalVolume = CompositeCriticalVolume +
    xMole(i) * CriticalVolume(i - 1) ^ (1# / 3#)
  Next i
  CompositeCriticalVolume = CompositeCriticalVolume ^ 3
End Function
Function CompositeCriticalPressure(xMole As Range) As
    Double
'this function returns the critical pressure in psia of
    the mixture given the mole fractions
  Dim MW As Double
  Dim Pc As Double
  Dim R As Double
  Dim Tc As Double
  Dim vc As Double
  Dim zc As Double
  MW = CompositeMolecularWeight(xMole)
  R = 1545.349 / MW
  Tc = CompositeCriticalTemperature(xMole)
  vc = CompositeCriticalVolume(xMole)
  zc = CompositeCriticalCompressibility(xMole)
  Pc = zc * R * Tc / vc / 144#
  CompositeCriticalPressure = Pc
End Function
```

```
Function RedlichKwongA(ByVal i As Integer) As Double
'this function returns Redlich-Kwong's coefficient a
   [English units] for the constituent
  Dim a As Double
  Dim MW As Double
  Dim Pc As Double
  Dim R As Double
  Dim Tc As Double
  MW = MolecularWeight(i)
  R = 1545.349 / MW
  Tc = CriticalTemperature(i)
  Pc = CriticalPressure(i)
  a = 0.4278 * R ^ 2 * Tc ^ 2.5 / (Pc * 144#)
  RedlichKwongA = a
End Function
Function RedlichKwongB(ByVal i As Integer) As Double
'this function returns Redlich-Kwong's coefficient b
   [English units] for the constituent
  Dim b As Double
  Dim MW As Double
  Dim Pc As Double
  Dim R As Double
  Dim Tc As Double
  MW = MolecularWeight(i)
  R = 1545.349 / MW
  Tc = CriticalTemperature(i)
  Pc = CriticalPressure(i)
  b = 0.0867 * R * Tc / (Pc * 144#)
  RedlichKwongB = b
End Function
Function CompositeRedlichKwongA(xMole As Range) As
   Double
'this function returns Redlich-Kwong's coefficient a
   [English Units] for the mixture
  Dim i As Integer
  CompositeRedlichKwongA = 0#
  For i = 1 To 21
    CompositeRedlichKwongA = CompositeRedlichKwongA +
    xMole(i) * RedlichKwongA(i - 1)
  Next i
End Function
Function CompositeRedlichKwongB(xMole As Range) As
   Double
'this function returns Redlich-Kwong's coefficient b
   [English Units] for the mixture
  Dim i As Integer
  CompositeRedlichKwongB = 0#
  For i = 1 To 21
```

```
        CompositeRedlichKwongB = CompositeRedlichKwongB +
        xMole(i) * RedlichKwongB(i - 1) ^ (1# / 3#)
        Next i
        CompositeRedlichKwongB = CompositeRedlichKwongB ^ 3
        End Function
```

R-K isotherms for typical natural gas composition are also calculated in the spreadsheet and illustrated in the following figure:

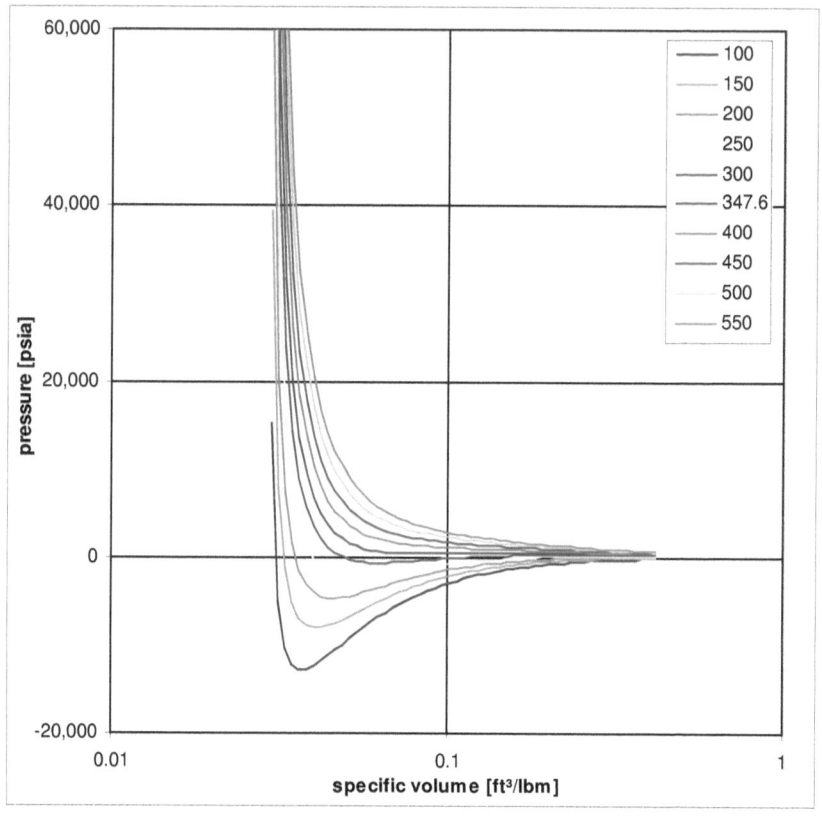

**Figure 24. Pressure vs. Specific Volume for Natural Gas**

Ammonia Reactions

Ammonia injection is often used to control NOx in the exhaust of gas turbines, but this may result in the creation of other unwanted products, especially if sulfur is present. This is why ammonia injection is recommended when burning low sulfur natural gas. I know of a case where operators switched from one fuel source to another, not realizing that sulfur in the new fuel would combine with the ammonia. The systems in question were not intended to

50

control any and all emissions—only to control the ones expected during the design phase. The fuel swap came after construction, long after the design was finalized. The offending emissions continued for some time, as no effort was made to monitor them—another disconnect between design and operation. As a huge lawsuit resulted from this mistake, I have scrambled the fuel composition to obfuscate the unintentional perpetrator.

The reaction in question is input by the following text file:

```
*a reacting mixture of S, O, C, H, and N
        !S 12O 4C 12H 6ON
              =!
Ammoacsfate Ammobicarb Ammocarb Ammonia Ammosfate
Cdiox Cmonox Cos Cs Ctriodia Cyanide H Hdia Hsulfide
Hydroxl Ndia Ndiox Nmonox Nndiox Nnmonox Nnpentaox
Nntetraox Nntriox Ntriox O Odia Sdiox Smonox Striox
              Sulfuricacid Water
```

As described in Appendix B, the first exclamation point (!) preceding all of the reactants and second (!) preceding all of the products activates real property calculations (i.e., non-ideal gas fugacities, compressibilities, and mixing rules). Real properties are deactivated by default, but can be activated for each reactant or product by following it by (!). A preceding (!) activates the entire set. The results at 1 atm. are:

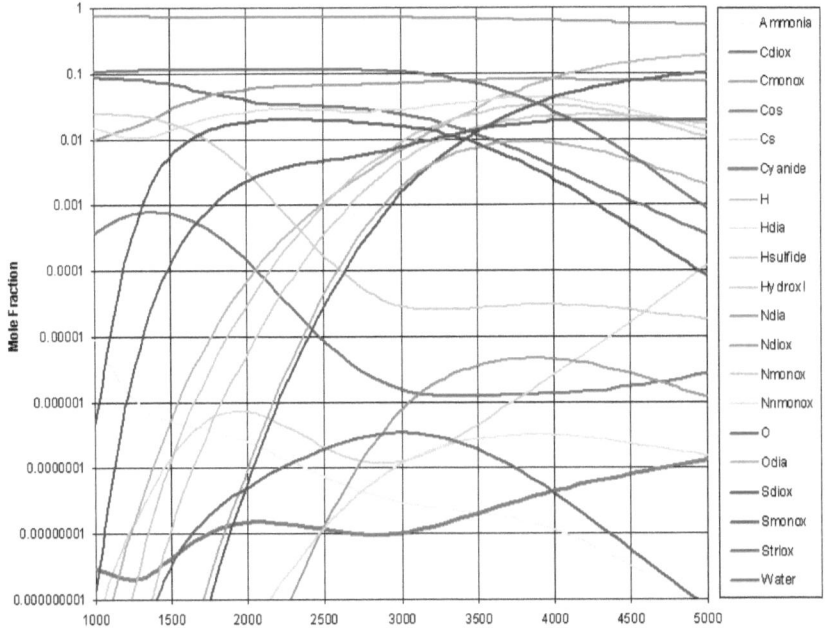

**Figure 25. Impact of Temperature on Combustion of Landfill Gas**

51

Cyanide is the brown curve (this color nicely corresponds to what "hit the fan" with the lawsuit). Notice the other nasties: CO, COS, $H_2S$, SO, $SO_2$, $SO_3$, and $H_2SO_4$! Erosion of equipment in contact with the exhaust was the first sign of trouble. A welder sent to "fix" the problem sounded the alert when he smelled rotten eggs.

## Chapter 8. Aqueous Reactions and Ions

Many important chemical reactions occur in solutions, that is liquids, rather than gases. While one might treat liquids as ideal gases and proceed to calculate product moles, this assumption is not likely to result in accurate predictions. We have already covered most of the equations needed to handle liquids and ions so that only a few things are still lacking. The first reaction we will consider is:

```
! 40Water Mgsulfide =!
   Water H Hdia H^ O Odia Oh~
 Mg Mg^^ Mghydroxide Mgsulfide Mgsulfate
 S Sdiox Smonox Striox Sulfate~~ Sulfite~~
       Bisulfate~~ Bisulfite~~
```

Again, the two leading (!)s activate non-ideal gas calculations for all reactants and products. Because plus (+) and minus (-) already mean something in equations and there are no subscript or superscript distinctions in a text file, we must use some other symbol to uniquely indicate positive and negative ions. I have chosen the carat (^) for positive charges and the tilde (~) for negative ones. Notice also that every chemical species begins with an upper case letter (A-Z) and optionally continues with lower case (a-z). Species names do not contain numbers (0-9), as there would be no way (without subscripts) to distinguish these from the number of moles. The results are listed below:

```
Tp=312 K, Pp=1 atm, Q/RTo=-50
substance state      Y          X         Z    F    H/RTo   S/R    G/RTo
-----------------------------------------------------------------------
Water        !lqd 40.0000000 0.9756098 0.10 0.97 -106.9  11.77  -129.6
Mgsulfide    !sol  1.0000000 0.0243902 0.01 1.12 -134.6  11.06  -155.9
-----------------------------------------------------------------------
reactants         41.0000000 1.0000000 0.09 0.97 -4410.8 481.76 -5340.9
=======================================================================
H            !gas  0.0000000 0.0000000 1.00 1.04   90.2  45.35     2.6
Hdia         !gas  0.9906086 0.0241556 1.00 1.04    3.2  19.17   -33.8
O            !gas  0.0000000 0.0000000 1.00 1.02  102.9  50.40     5.6
Odia         !gas  0.0000000 0.0000000 1.00 1.02    3.4  56.40  -105.5
Sdiox        !gas  0.0000004 0.0000000 1.00 0.97 -114.7  48.56  -208.5
Smonox       !gas  0.0000000 0.0000000 1.00 0.98    5.3  57.04  -104.8
Striox       !gas  0.0000000 0.0000000 1.00 0.96 -153.9  62.14  -273.9
Water        !lqd 38.0181256 0.9270575 0.10 0.97 -106.9  11.81  -129.7
Bisulfate~~  !aqu  0.0000000 0.0000000 1.00 0.62 -545.3  47.39  -636.8
Bisulfite~~  !aqu  0.0000000 0.0000000 1.00 0.65 -308.2  33.55  -372.9
H^           !aqu  0.0007476 0.0000182 1.00 1.04    0.5   9.03   -17.0
Mg^^         !aqu  0.0090834 0.0002215 1.00 1.01 -187.8 -10.44  -167.6
Oh~          !aqu  0.0000014 0.0000000 1.00 1.02 -109.4   1.87  -113.0
Sulfate~~    !aqu  0.0000001 0.0000000 1.00 0.90 -366.2  17.75  -400.5
Sulfite~~    !aqu  0.0000000 0.0000000 1.00 0.93 -255.8  23.17  -300.6
Mg           !sol  0.0000000 0.0000000 0.01 1.12    2.7  38.27   -71.2
Mghydroxide  !sol  0.9908913 0.0241625 0.01 1.14 -364.3  15.18  -393.6
Mgsulfate    !sol  0.0000223 0.0000005 0.01 1.09 -507.5  31.49  -568.3
Mgsulfide    !sol  0.0000030 0.0000001 0.01 1.15 -134.6  25.72  -184.2
S            !sol  0.9999741 0.0243840 0.01 0.82   -8.4   3.96   -16.1
-----------------------------------------------------------------------
products          41.0094579 1.0000000 0.97 0.97 -4432.2 487.01 -5372.5
-----------------------------------------------------------------------
difference         0.0094579 0.0000000 0.00 0.00  -21.4   5.25   -31.5
Note: the ! above indicates non-ideal behavior
```

The products have been sorted by state: gas, liquid, aqueous ion, and solid. From thermodynamics alone, we can't tell whether a gas will be dissolved (intimate contact with the solution) or evolved (bubble up and leave the solution) or whether a solid will be dissolved (intimate contact with the solution), suspended (minimal contact with the solution), or precipitated (no appreciable contact with the solution). Of course, all these products must be present while the reaction is occurring for our analysis to be valid.

The compressibility (Z) for liquids defaults to 0.10, gases and ions 1.00, solids 0.01. Note that the fugacity coefficient (F) varies. If we were to base our calculations on ideal gas assumptions, significant errors would be expected. If we were to merely correct for compressibility, we would greatly underestimate the activity of the ions for certain and perhaps the liquids and solids. The partial pressures exerted by these components with Z<<1 might be quite small, but they participate in the reaction at a much higher level, which is captured by the fugacity coefficients.

In the preceding table, we see that almost all of the MgS has converted to $Mg(OH)_2$, forming $H_2$ gas (which evolves), and S (which precipitates). Anyone who has "played" with magnesium and its compounds has witnessed this reaction. We would not see this same outcome when assuming ideal gas behavior. Because "we" know this outcome is correct (from personal experience), we know the real fluid calculations are working correctly.

We will next replace S with P and see that the outcome changes (because sulfur, phosphorous, and their compounds have different properties).

```
! 40Water Mgphosphate =! H Hdia O Odia Water
   Hhphosphate~ Hphosphate~~ H^ Mg^^ Oh~ Phosphate~~~
            Mg Mghydroxide Mgphosphate P
```

```
Tp=312 K, Pp=1 atm, Q/RTo=0
substance    state      Y          X        Z    F     H/RTo    S/R     G/RTo
------------------------------------------------------------------------------
Water        !lqd 40.0000000 0.9756098 0.97 0.97  -106.9   11.77  -129.6
Mgphosphate  !sol  1.0000000 0.0243902 1.00 0.96 -1501.0   40.27 -1578.7
------------------------------------------------------------------------------
reactants         41.0000000 1.0000000 0.97 0.97 -5777.2  510.96 -6763.7
==============================================================================
H            !gas  0.0000000 0.0000000 1.00 1.04    90.2   45.36     2.6
Hdia         !gas  0.0189833 0.0004632 1.00 1.04     3.2   23.29   -41.7
O            !gas  0.0000000 0.0000000 1.00 1.02   102.9   50.41     5.6
Odia         !gas  0.0000000 0.0000000 1.00 1.02     3.4   56.40  -105.5
Water        !lqd 39.8858112 0.9732749 0.10 0.97  -106.9   11.77  -129.6
Hhphosphate~ !aqu  0.0380577 0.0009287 0.10 0.82  -526.4   15.72  -556.8
Hphosphate~~ !aqu  0.0000043 0.0000001 0.10 0.90  -520.6    8.05  -536.1
H^           !aqu  0.0001025 0.0000025 0.10 1.04     0.5   11.09   -20.9
Mg^^         !aqu  0.0000040 0.0000001 0.10 1.01  -187.8   -2.62  -182.7
Oh~          !aqu  0.0000107 0.0000003 0.10 1.02  -109.4   -0.22  -108.9
Phosphate~~~ !aqu  0.0000000 0.0000000 0.10 0.90  -514.6   -0.79  -513.1
Mg           !sol  0.0000000 0.0000000 0.01 1.16     2.7   39.49   -73.5
Mghydroxide  !sol  0.0570890 0.0013931 0.01 1.21  -364.3   18.86  -400.8
Mgphosphate  !sol  0.9809690 0.0239371 0.01 0.97 -1501.0   40.31 -1578.8
P            !sol  0.0000000 0.0000000 0.01 1.15     2.3   37.74   -70.6
------------------------------------------------------------------------------
products          40.9810318 1.0000000 0.97 0.97 -5777.2  511.10 -6764.0
------------------------------------------------------------------------------
difference        -0.0189682 0.0000000 0.00 0.00     0.0    0.13    -0.3
Note: the ! above indicates non-ideal behavior
```

At the same conditions (312°K/38°C and 1 atm.) and with the differing properties of P vs. S, we see that most of the initial $Mg_3(PO_4)_2$ remains and very little $H_2$ is formed. The pH of the two solutions will also be quite different. In the first reaction, virtually all of the S ended up as a precipitate, while in the second reaction, none of the P precipitates.

### Dissolution of Transite

My motivation to purse analysis of aqueous reactions was failure of transite (asbestos fiber reinforced cement materials). We now consider that reaction:

```
! 1000Water 0.5Cdiox 0.00001Odia 0.1Tricaaloxide
 0.1Casilicateb 0.1Cawollastoni 0.1Mgchrysotile
 0.005Ca^^ 0.01Clmonox~ =! Water Cldia Cdiox Hdia
   Odia Algibbsite Tricaaloxide Calcite Caoxide
 Casilicateb Cawollastoni Mghydroxide Mgchrysotile
      Sioquartz H^ Al^^^ Ca^^ Oh~
    Cl~ Clmonox~ Ctriox~~ Hcarbonate~ Mg^^
```

Tp=576 K, Pp=1 atm, Q/RTo=-63

| substance | state | Y | X | Z | F | H/RTo | S/R | G/RTo |
|---|---|---|---|---|---|---|---|---|
| Water | !lqd | 99.00000 | 0.990842 | 0.10 | 0.97 | -106.9 | 11.75 | -129.6 |
| Cdiox | !gas | 0.50000 | 0.005004 | 1.00 | 1.00 | -154.0 | 31.60 | -215.0 |
| Odia | !gas | 0.00001 | 0.000000 | 1.00 | 1.02 | 3.4 | 40.82 | -75.4 |
| Ca^^ | !aqu | 0.00500 | 0.000050 | 0.10 | 0.99 | -218.4 | 1.03 | -220.4 |
| Clmonox~ | !aqu | 0.01000 | 0.000100 | 0.10 | 0.97 | -42.7 | 11.60 | -65.2 |
| Tricaaloxide | !sol | 0.10000 | 0.001000 | 0.01 | 0.94 | -1423.6 | 44.88 | -1510.2 |
| Casilicateb | !sol | 0.10000 | 0.001000 | 0.01 | 1.06 | -916.3 | 29.98 | -974.2 |
| Cawollastoni | !sol | 0.10000 | 0.001000 | 0.01 | 1.14 | -649.9 | 21.65 | -691.7 |
| Mgchrysotile | !sol | 0.10000 | 0.001000 | 0.01 | 0.94 | -1730.1 | 51.78 | -1830.0 |
| reactants | | 99.91501 | 1.000000 | 0.97 | 0.97 | -11134.2 | 1194.35 | -13440.2 |
| Water | !lqd | 98.59497 | 0.986882 | 0.10 | 0.97 | -106.9 | 11.76 | -129.6 |
| Cldia | !gas | 0.00000 | 0.000000 | 1.00 | 0.97 | 3.8 | 58.22 | -108.7 |
| Cdiox | !gas | 0.14319 | 0.001433 | 1.00 | 1.00 | -154.0 | 32.84 | -217.4 |
| Hdia | !gas | 0.00002 | 0.000000 | 1.00 | 1.04 | 3.2 | 30.94 | -56.5 |
| Odia | !gas | 0.00000 | 0.000000 | 1.00 | 1.02 | 3.4 | 57.25 | -107.1 |
| H^ | !aqu | 0.00000 | 0.000000 | 0.10 | 1.04 | 0.5 | 14.91 | -28.3 |
| Al^^^ | !aqu | 0.00000 | 0.000000 | 0.10 | 1.01 | -213.8 | -8.58 | -197.2 |
| Ca^^ | !aqu | 0.00011 | 0.000001 | 0.10 | 0.99 | -218.4 | 4.73 | -227.6 |
| Oh~ | !aqu | 0.00109 | 0.000011 | 0.10 | 1.02 | -109.4 | -4.07 | -101.5 |
| Cl~ | !aqu | 0.01000 | 0.000100 | 0.10 | 0.97 | -66.9 | 13.40 | -92.8 |
| Clmonox~ | !aqu | 0.00000 | 0.000000 | 0.10 | 0.97 | -42.7 | 33.98 | -108.3 |
| Ctriox~~ | !aqu | 0.00000 | 0.000000 | 0.10 | 0.96 | -272.6 | 9.29 | -290.5 |
| Hcarbonate~ | !aqu | 0.00924 | 0.000092 | 0.10 | 0.91 | -282.0 | 18.95 | -318.6 |
| Mg^^ | !aqu | 0.00000 | 0.000000 | 0.10 | 1.01 | -187.8 | 3.90 | -195.3 |
| Algibbsite | !sol | 0.10000 | 0.001000 | 0.01 | 1.05 | -1012.8 | 35.99 | -1082.2 |
| Tricaaloxide | !sol | 0.00000 | 0.000000 | 0.01 | 0.92 | -1423.6 | 68.02 | -1554.9 |
| Calcite | !sol | 0.34756 | 0.003478 | 0.01 | 1.09 | -477.7 | 21.04 | -518.3 |
| Caoxide | !sol | 0.00000 | 0.000000 | 0.01 | 1.19 | -251.5 | 25.71 | -301.2 |
| Casilicateb | !sol | 0.00003 | 0.000000 | 0.01 | 1.03 | -916.3 | 38.12 | -989.9 |
| Cawollastoni | !sol | 0.25724 | 0.002574 | 0.01 | 1.11 | -649.9 | 20.40 | -689.3 |
| Mghydroxide | !sol | 0.29947 | 0.002997 | 0.01 | 1.16 | -364.3 | 17.65 | -398.4 |
| Mgchrysotile | !sol | 0.00017 | 0.000001 | 0.01 | 0.92 | -1730.1 | 57.48 | -1841.0 |
| Sioquartz | !sol | 0.14237 | 0.001425 | 0.01 | 1.19 | -362.5 | 13.59 | -388.7 |
| products | | 99.90552 | 1.000000 | 0.97 | 0.97 | -11161.4 | 1187.59 | -13454.3 |
| difference | | -0.00948 | 0.000000 | 0.00 | 0.00 | -27.2 | -6.76 | -14.2 |

Note: the ! above indicates non-ideal behavior

This thermodynamic analysis fits with large-scale observations (i.e., deterioration in the field) as well as laboratory tests, including ones intended to accelerate the dissolution process by increasing temperature. CO2 is produced and also Calcite, which is somewhat soluble. The reaction also produces $Mg(OH)_2$, which is soluble and the most critical path of material failure. The range of temperatures for cooling water can't extend beyond 0°C to 100°C (freezing to boiling), so we will calculate and plot the moles over this range:

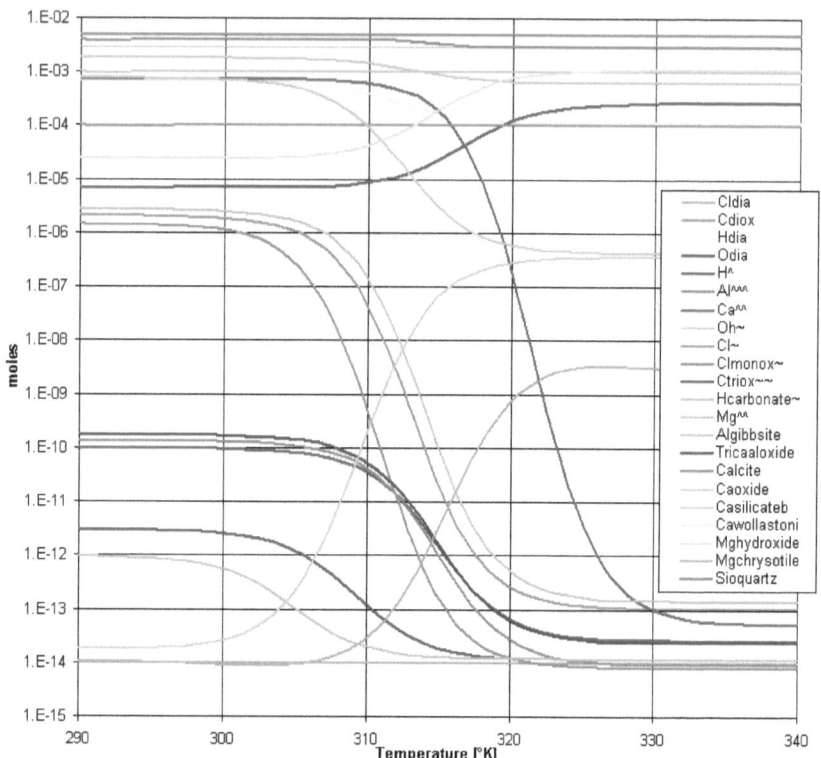

**Figure 26. Impact of Temperature on Dissolution of Transite**

In this figure we see that the impact of temperature is most pronounced over the range 300°K to 320°K, which is 37°C to 47°C (or 90°F to 116°F). Unfortunately for the power industry, this is the normal operating range for cooling towers, where the materials are used. The situation could hardly be worse for operation and maintenance. Anecdotal evidence pointed to this conclusion, but the software wasn't yet operational and wouldn't be until after decisions were made and other alternatives explored. Changes in cement, fibers, and manufacturing practice were pursued in an effort to address this critical material failure. Several years after this effort ended, the software finally attained the capability of properly analyzing this complex reaction and the results matched expectation. The wait was over...

## The Secret Revealed

Before we continue with more examples, *it's time for the big reveal...* The reason I pursued this topic to the point of writing software, while stranded in the Chicago airport, was the failure of fiber-reinforced cement materials due to chemical attack in an aqueous solution—specifically transite and similar materials in cooling water systems. Molecules essential to the integrity of the materials were being preferentially leached out into the water, resulting in structural failures. Replacement of the original asbestos-based fibers with environmentally acceptable alternatives proved to be even more problematic. In short, the approved alternatives broke down even faster and failed more dramatically than the original banned materials. This problem was of great financial concern to the power industry as well as others using similar products.

I was able to develop the software and to solve various reactions, such as combustion, obtaining good agreement with experimental data as well as analytical studies available in the open literature. When I applied the same analysis to aqueous solutions, problems arose immediately. I traced these back to the source: an ill-conditioned matrix A, the Hessian, containing second partial derivatives of the total Gibbs free energy, $G=\Sigma y_I g_I$, with respect to the molar abundances, $y_I$.

Realizing that aqueous solutions aren't ideal gases, I had derived the corrections described in Chapter 5 and applied them to matrices A and B. Matrix B contains the product free energies, $-g_I$. That wasn't the source of this problem. The terms of A weren't zero or even close to it. I mention on page 29 that the diagonal terms of A are much larger than the off-diagonal ones. That was no longer true! I added an estimation of the condition number to the Gauss elimination routine used to solve the simultaneous linear equations at each step of the steepest descent.[21] Comparing the condition numbers for the reactions that converged to those that didn't revealed the problem: increasingly ill-conditioned matrix A.

After further experimentation, I found that this problem only occurred if the fugacity coefficient, $\varphi$, or the compressibility, Z, were less than unity. Reduced fugacities are to be expected in aqueous vs. gaseous reactions. Reduced compressibilities are a given in liquids, with Z typically less than 0.1. I tried all

---

[21] Definitions of the condition number of a matrix vary, but share one thing in common: if this number is very large (by one definition or very small by another), the matrix becomes singular. After reducing the matrix to upper-triangular form, which is part of the process of Gauss elimination, the product of the diagonal elements indicates the condition number.

sorts of matrix techniques and *high accuracy* solution algorithms.[22] I explored damping factors, step-length algorithms, hybrid directional searches, and everything else I could find in the literature. Nothing worked. The problem was still there: ill-conditioned matrix A. Eventually, funding ran out and I moved on to other things, leaving the problem unsolved.

Several years later I was working on a totally different project: analysis of subterranean aquifers by pulsed pumping of groundwater from sampling wells.[23] I was also solving a constrained minimization problem for this project: finding the hydraulic properties that best explained the response of each well (i.e., lowest residual error between data and calculated result). The Method of Steepest Descent was the obvious approach, but the partial derivatives could not be calculated analytically (due to factors arising from and limitations of field data that was extremely expensive to obtain). Because of this limitation, I began using the derivative-free approach, developed by Broyden.[24]

This approach worked fairly well for the problem at hand, but was slow to converge and often diverged if the initial estimates were too far off or the data was too noisy or consisted of insufficient points. This led me to develop several modifications to the original Broyden method, which I eventually published.[25] The heart of Broyden's method is a rank-one update to the Hessian.

Rank-one means a single vector (one row or one column), which spans only one dimension of the space defined by the complete matrix. The vector (Nx1) times it's transpose (1xN) results in a square (NxN), which is added to the full (NxN) matrix. The goal in Broyden's method (and its modifications) is to determine which rank-one update that can be deduced from the latest computational effort, should be added to the previous Hessian, forming an updated Hessian, which will ultimately lead to the minimum by the iterative process.

The answer is: the one that maximizes the condition number of the Hessian. As I programmed this concept into the code I was using to solve the groundwater project, the corresponding unsolved problem of analyzing aqueous chemical reactions hit me. Was there something I could add to that Hessian, which would rescue the condition number from singularity? Yes there is!

I tried several things until I found one that worked. First, I wanted to add something that wouldn't change the solution. There's no point finding a wrong solution. This narrowed down the search to rank-one updates that don't impact

---

[22] I couldn't help wondering at the time, "Who would want a *low* accuracy solution?"

[23] I'm not a geohydrologist, but I am an applied mathematician. I volunteered to support this work because I was the only one in the room with the necessary math skills.

[24] Broyden, C., "A New Method of Solving Nonlinear Simultaneous Equations," Computational Journal, Vol. 12, pp. 94 99, 1969.

[25] Benton, D. J., "Applications of a Hybrid Derivative-Free Algorithm for Locating Extrema," SIAM SEAS, 1991.

the final result, although they will most likely alter the intermediate solutions along the iterative path toward the final result. At first, I tried many complicated things. After all, a tough problem must have a complicated solution... right?

My first attempts were fruitless. I even tried random numbers. During this process I discovered that increasing the diagonal terms of A by values significantly greater than unity resulted in extraneous solutions. Increments significantly less than unity failed to impact the outcome. I also found that normalizing the rows (something that is sometimes done in matrix solution algorithms) proved disastrous.

Once by accident, I entered a complicated equation that worked like a charm! It was an act of desperation, but it worked. I had the solution... now to figure out what it meant and why it worked. Through algebraic manipulation, the complicated formula reduced down to simply: $y_I = y_I$. Could it possibly be that simple? Of course, adding this most simple rank-one update impacts the solution at every step, because $y_I$ changes as we approach the solution. When the minimum is reached, there is no impact on the solution of adding this and Gauss elimination removes it one step at a time when pivoting.

$$y_I = y_I$$
*That's it!*
**This works even with tiny values of $\varphi$ and $Z$.**

## Chapter 9. Air:Fuel Ratio

Optimum air:fuel ratio is very important to the operation of any combustion engine. Introductory presentations of chemical reactions and combustion consider only ideal reactions. Only very advanced courses ever cover real reactions, including dissociation. Selection of air:fuel ratio is often anecdotal, but can be analyzed thermodynamically, which we will now demonstrate. The first reaction to consider will be combustion of hydrogen and oxygen at 1 atm. and 2000°K. The following graph shows the product mole fractions as a function of the moles of $O_2$ provided:

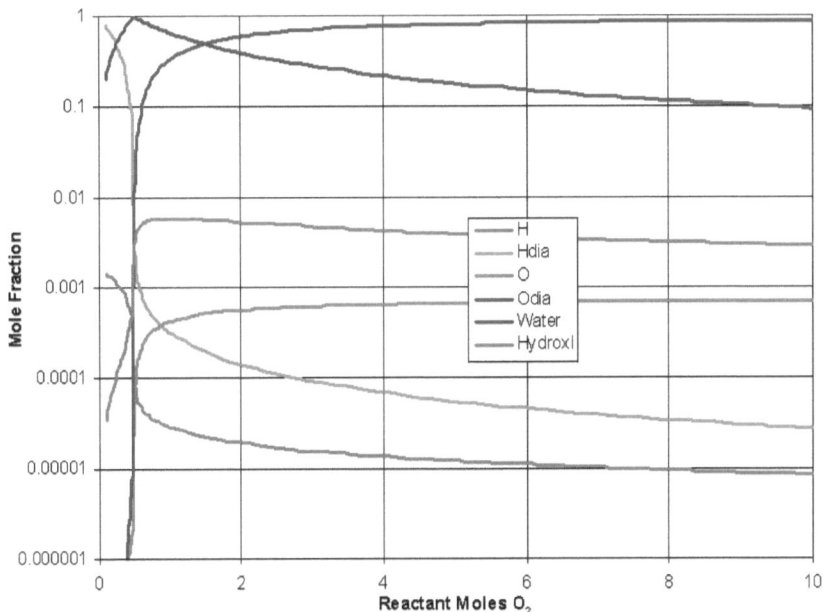

**Figure 27. Impact of $O_2$ on the Combustion of $H_2$**

There is a very sharp change in all products at 0.5, which is the stoichiometric ratio.[26] This is also the highest point in the violet $H_2O$ curve, which means the greatest net release of energy. All other products besides $H_2O$ require energy. For that reason alone we want to produce as little of them as possible. Notice how even the dissociated oxygen (O) increases with the number of reactant moles of $O_2$. At least the hydroxl (OH) diminishes with increasing $O_2$. This abrupt behavior is typical for combustion reactions so that the science corresponds to the common sense and reinforces the importance of such analyses.

---

[26] There is exactly enough oxygen ($O_2$) to combine with the fuel ($C_nH_m$) in an ideal reaction, producing only carbon dioxide ($CO_2$) and water vapor ($H_2O$)

This next figure is the same data, only plotted on part of the X-axis. Note how sharply the transition occurs.

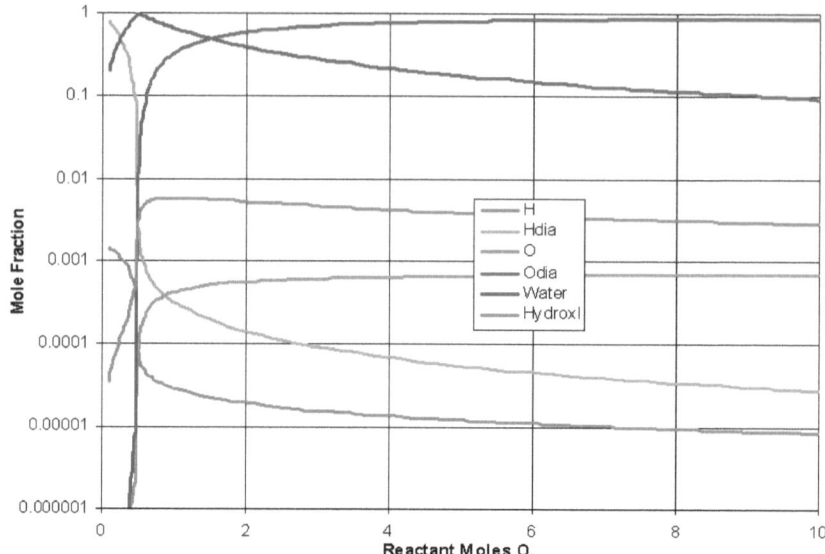

**Figure 28. Impact of O$_2$ on the Combustion of H$_2$ (detail)**

We next consider methane and oxygen.

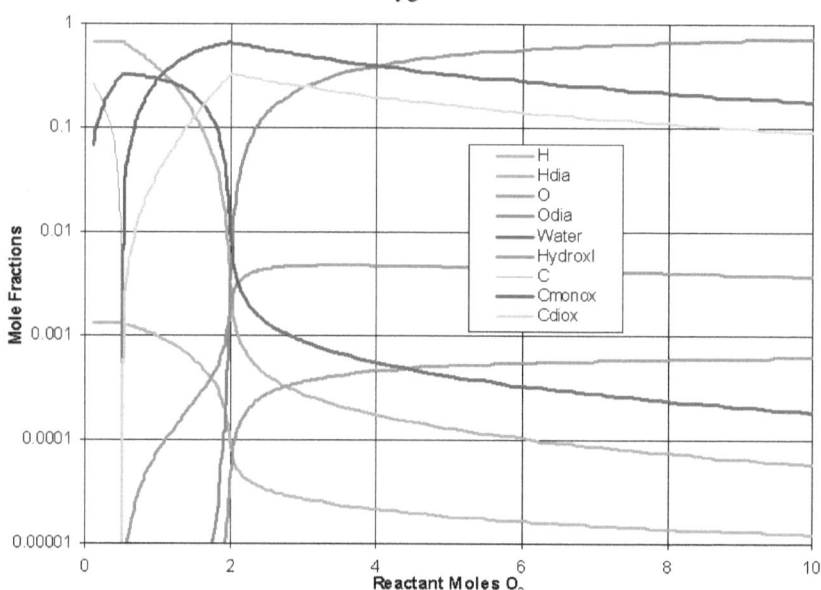

**Figure 29. Impact of O$_2$ on the Combustion of Methane**

In this reaction we see two transitional points: one at 0.5 and a second at 2. The first is the same as before, dominated by hydrogen, and the second is the stoichiometric ratio for methane. Again, the maximum $H_2O$ (violet curve) and $CO_2$ (sky blue curve) both occur at 2 reactant moles of $O_2$. This is also the point where we waste the least energy producing unwanted dissociation products. Experience and science line up again. The same data on over a smaller range:

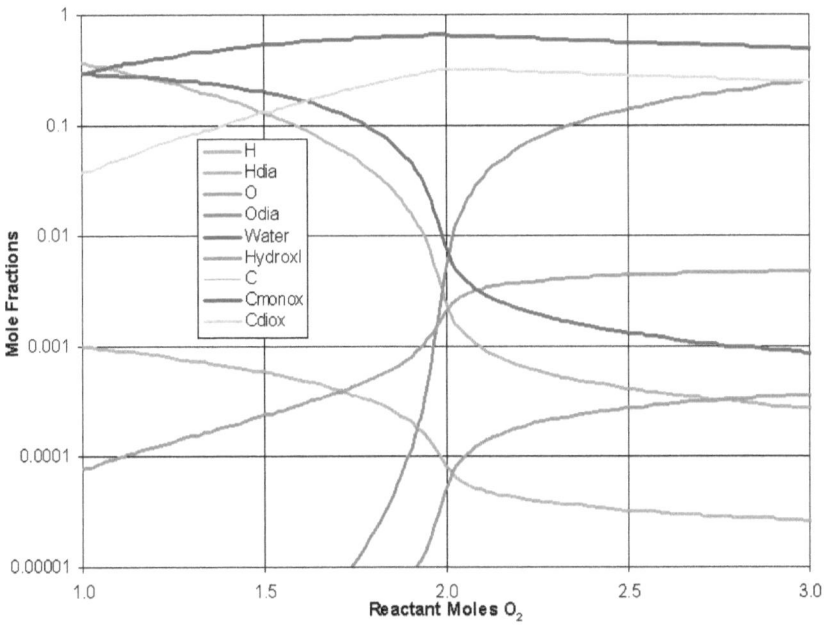

**Figure 30. Impact of O2 on the Combustion of Methane (detail)**

*It should be abundantly clear by this point how important it is to have a program designed to process this information (databases and reactions) and solve these complex equations efficiently, even stepping through loops of different variables and generating multiple plots.*

*Best of all... it's free!*

We next consider combustion of natural gas and air at 2000°K and 1 atm.(same hydrocarbon mixture as on page 34).

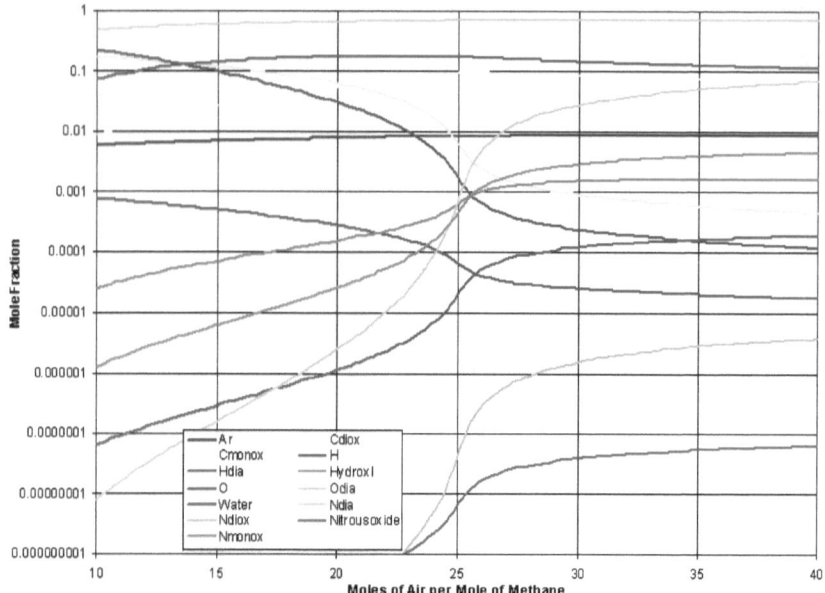

**Figure 31. Impact of Air/Fuel Ratio on the Combustion of Natural Gas**

We see that the stoichiometric ration for this fuel is about 26. The brown $H_2O$ curve peak is very flat, maximizing out at about 22.5. The green $CO_2$ curve peaks at about 21. As before, the unwanted products increase with excess air (i.e., more than stoichiometric). This is one reason why gas turbines divert most of the air past the combustor. Not only does this practice maximize the net energy released from the fuel, but also minimize the undesirable emissions.

## Chapter 10. H:C Ratio

In the previous chapter we considered air:fuel ratio with the same fuel composition. We will now consider H:C ratio, that is, different fuel compositions with the same air, though not the same air:fuel ratio. We will maintain the stoichiometric ratio while changing the fuel average hydrocarbon chain length. We will hold the combustion temperature constant at 2000°K and the pressure at 1 atm. Beginning with just $O_2$, we'll use a trick[27] to vary H:C while maintaining stoichiometry.

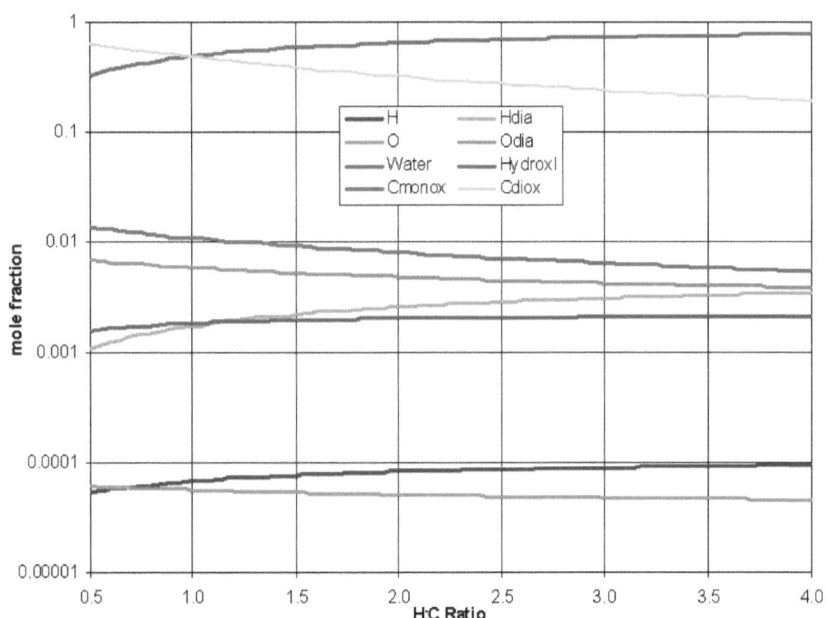

**Figure 32. Impact of H:C Ratio on Combustion Products with $O_2$**

Unlike the air:fuel ratio graphs in the last chapter, H:C ratio graphs don't exhibit maximums and minimums, only tapering trends. Not surprisingly, the moles of $H_2O$ (violet curve) gradually increase with H:C and the moles of $CO_2$ (sky blue curve) gradually decrease. Of course, the OH and CO follow corresponding trends, as these are both limited by the initial H:C, which is being varied along the X-axis. Note that dissociated O and diatomic $O_2$ both decrease with increasing H:C ratio, while unburned $H_2$ increases. All of these trends are intuitive so that the science follows intuition.

---

[27] We create two pseudo-compounds in the database: one a combination of hydrogen and oxygen and a second comprised of carbon and oxygen, neither of which has already combusted. We will do the same thing for hydrogen plus air and carbon plus air.

We will next consider $C_nH_m$ combustion with air, using the same database trick. Since we're creating these two fictitious hydrocarbons, we can't discern heating value, as that would becomes an arbitrary input, rather than an outcome. No matter, as we are fixing the temperature of the products. Eliminating all of the products less than one part per million, leaves:

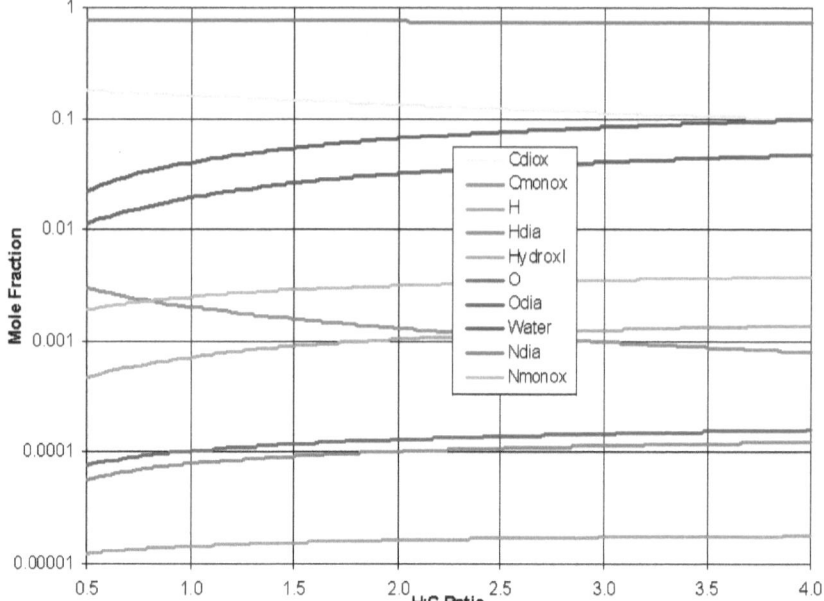

**Figure 33. Impact of H:C Ratio on Combustion Products with Air**

In this figure we do not see any critical point or maxima, which is consistent with the previous figure and, like it, distinct from the air:fuel ratio graphs of the previous chapter. If we eliminate all but CO and NOx:

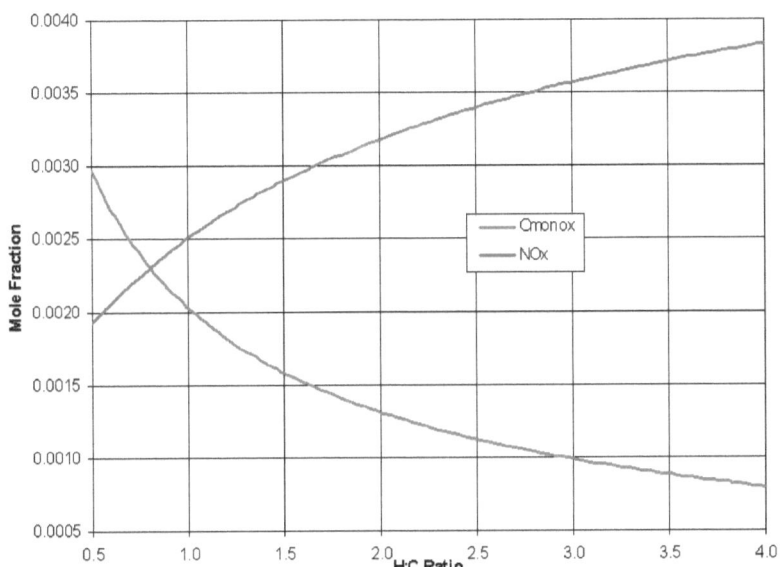

**Figure 34. Impact of H:C Ratio on Formation of CO and NOx**

There is also a continuous trend in OH:H, which is not surprising.

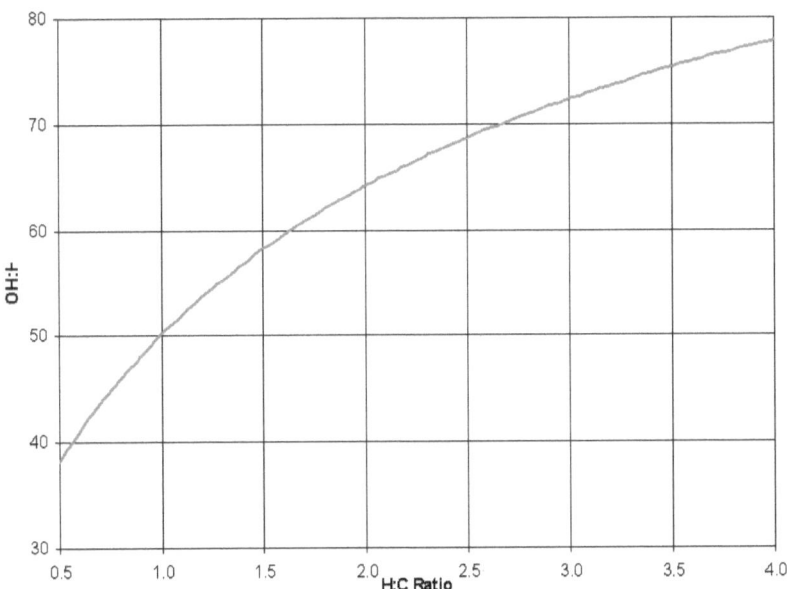

**Figure 35. Impact of H:C Ratio on OH:H Ratio**

# Chapter 11. Ammonia and NOx

Ammonia or urea is sometimes used to control NOx emissions in gas turbines and stationary diesel engines. This is most often injected into a selective catalytic reduction (SCR) reactor in the exhaust stream at essentially one atmosphere. Exhaust temperatures vary from 400°K to 700°K. We already saw what happens when there's sulfur in the fuel, so we won't repeat that here. Instead, we will begin with the simplest possible combination: hydrogen, oxygen, and ammonia. The reaction is:

```
!Ammonia+Hdia+0.5Odia=H+Hdia+Hydroxl+
O+Odia+Ozone+Water+Ammonia+N+Ndia+Ndiox
+Nitrousoxide+Nmonox+Nndiox+Nnpentaox
+Nntetraox+Nntriox+Ntriox
```

We maintain stoichiometric oxygen, maintain 2000°K, and 1 atm., while increasing NH$_3$ moles from 0.001 to 1. Moles of NOx are summed in the spreadsheet after solving the reaction.

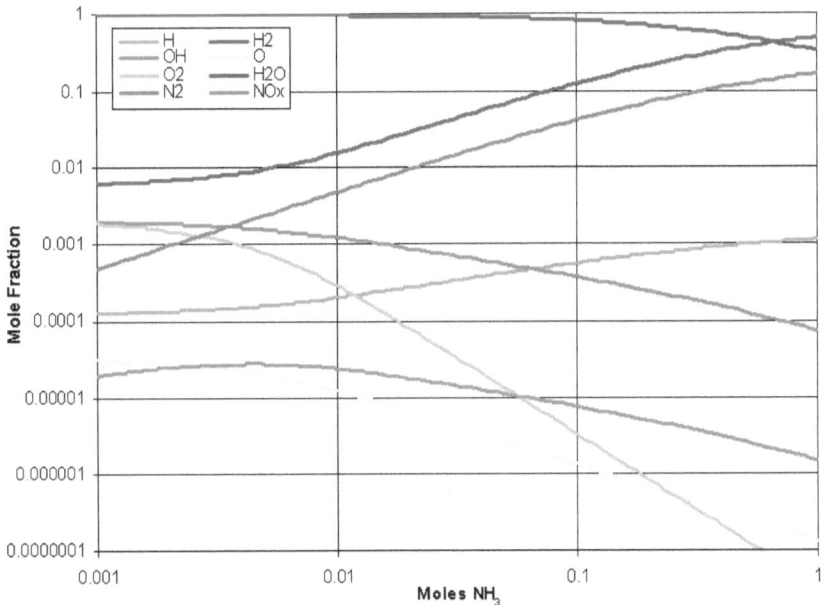

**Figure 36. Impact of Ammonia on Combustion of Hydrogen**

We see that NOx increases from 0.001 to 0.005, because there would be no nitrogen in the reactants if not for the NH$_3$. This is an artifact of the simulation and not a practical concern. Nobody burns hydrogen in a gas turbine. Indeed, the NOx decreases (red curve), as we would hope. Unfortunately, the H$_2$O decreases, as some of the O$_2$ combines with N to form NOx.

In this next figure we see the impact of ammonia on OH:H ratio:

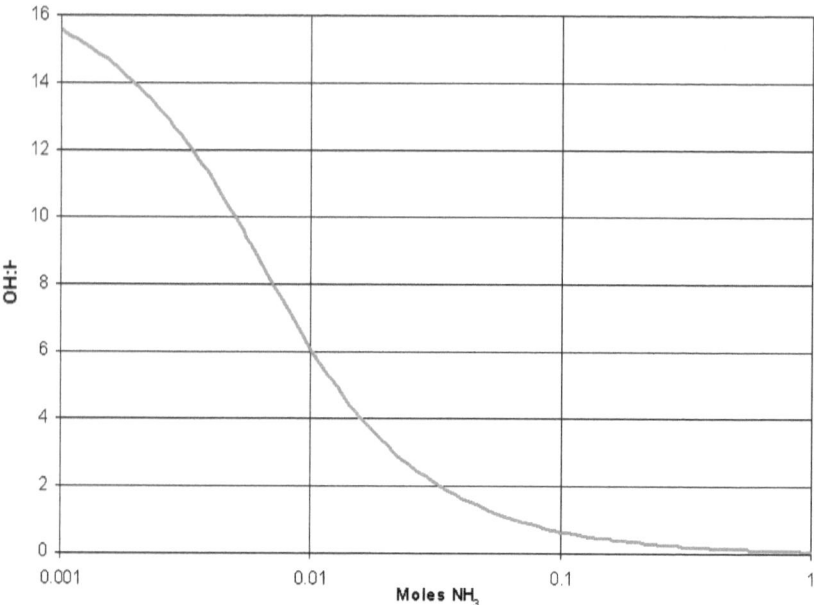

**Figure 37. Impact of Ammonia on OH:H Ratio ($H_2$)**

This trend is also to be expected when introducing ammonia and must be considered when designing systems. When we next consider methane as the fuel, we will provide some extra oxygen. In practice we always try to stay slightly above stoichiometric. The reaction is:

```
!Ammonia+Methane+3.5Odia=!H+Hdia+O+Odia
  +Water+Hydroxl+C+Cmonox+Cdiox+Ozone
    +Ammonia+N+Ndia+Ndiox+Nitrousoxide
       +Nmonox+Nndiox+Nnpentaox
         +Nntetraox+Nntriox+Ntriox
```

The results are illustrated in this next figure:

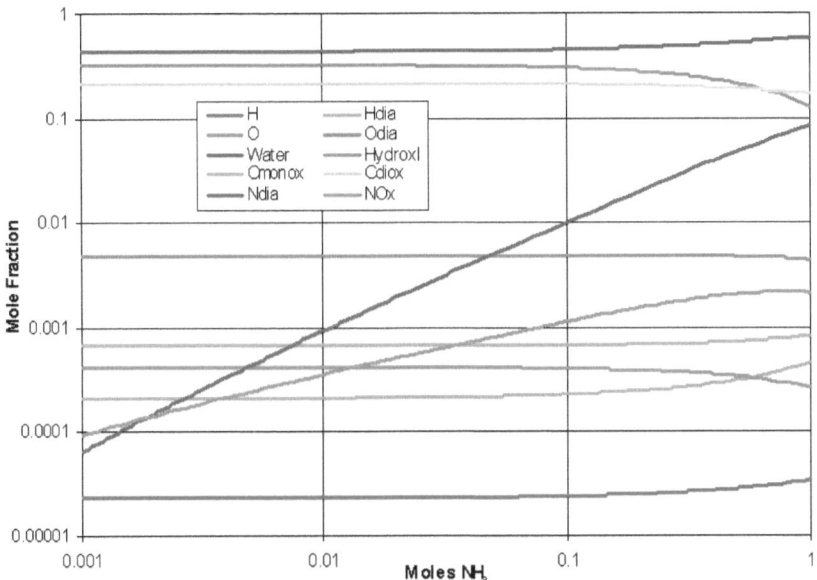

**Figure 38. Impact of Ammonia on the Combustion of Natural Gas**

We see that the mole fraction of $H_2O$ (violet curve) steadily increases, while the mole fraction of $CO_2$ slightly decreases. This isn't a fuel trend, this is an artifact of adding more hydrogen with the $NH_3$. Of course, the $N_2$ steadily increases, as expected. The unburned hydrogen (magenta $H_2$ and brown H curve) slowly rises, which is not desirable. The carbon monoxide (orange curve CO curve) also rises gradually. Most surprisingly, the mole fraction of NOx (red curve) increases over the entire range of $NH_3$ moles.

There are two things to consider in light of this result (increasing NOx with $NH_3$ injection for this particular reaction). First, we're only considering $O_2$ and not air, which is 78% nitrogen. Second, we rarely burn pure methane. Most natural gas has some $N_2$ in it, so we must deal with the nitrogen compounds regardless.

The OH:H curve for this reaction would be alarming, if we were to actually operate a system with this fuel and pure oxygen with ammonia injection, which would be foolish.

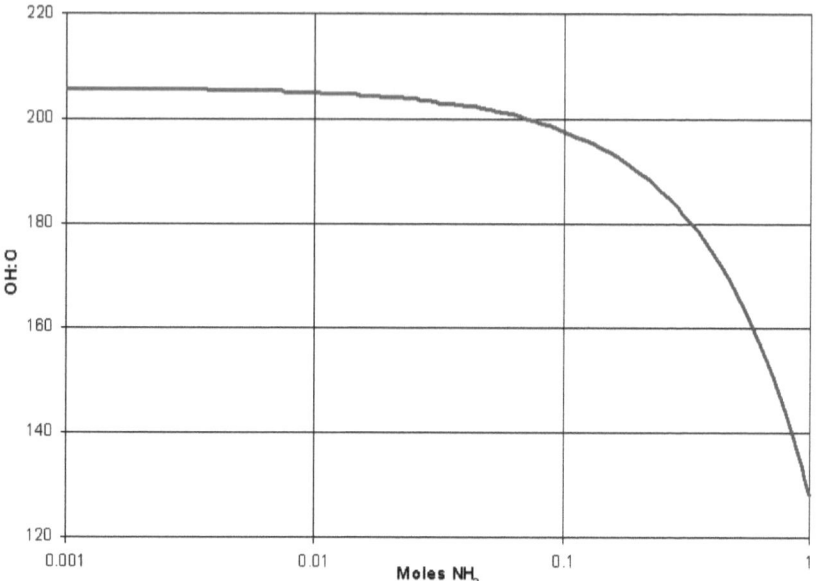

**Figure 39. Impact of Ammonia on OH:H Ratio (Natural Gas)**

We will next consider natural gas and slightly excess air: The reaction is:

```
!Ammonia+35Air+0.7578125 Methane+0.125 Ethane
+0.0625 Propane+0.03125 Butane+0.015625 Pentane
+0.0078125 Hexane=!Ar+C+Cdiox+Cmonox+H+Hdia+Hydroxl
+O+Odia+Ozone+Water+Ammonia+N+Ndia+Ndiox+Nitrousoxide
+Nmonox+Nndiox+Nnpentaox+Nntetraox+Nntriox+Ntriox
```

The results are shown in the following figure:

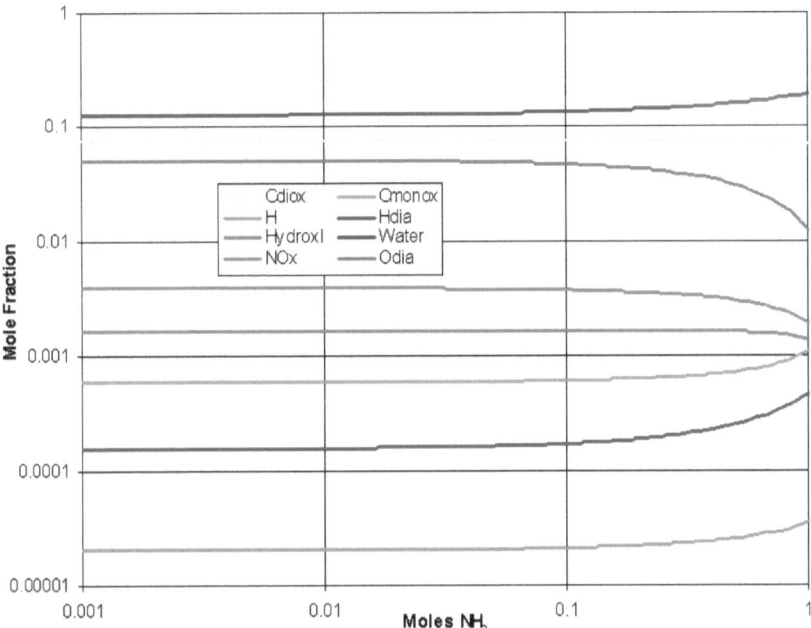

**Figure 40. Impact of NH$_3$ on Combustion of NG (with Excess Air)**

Here with what is a much more realistic situation (natural gas with air), we see that the NOx mole fraction (red curve) does decrease with increasing NH$_3$. However, the CO mole fraction goes up, which means that we must supply even more excess air to assure complete combustion. Note in this reaction that sufficient air was provided (i.e., there was still 5% remaining O2 (green curve) over the entire range of NH$_3$), but this didn't react with all of the H (magenta curve) or CO (orange curve).

The OH:H ratio is also much more manageable and trending downward.

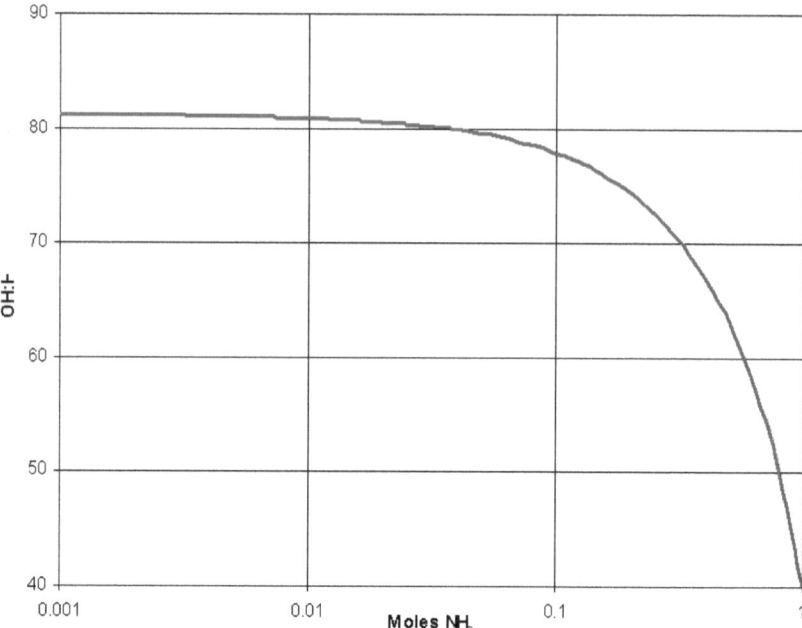

**Figure 41. Impact of NH3 on OH:H Ratio (of NG with Excess Air)**

## Chapter 12. Limestone and SOx

Low sulfur coal was at one time considered of little value because its heating value is typically less than high sulfur coal. This disparity in heating value does not follow from the sulfur content. While sulfur does burn, it doesn't release nearly as much energy per unit mass as carbon. The sulfur content and heating value of various coals arises from their origin: different plants/organisms and burial conditions. Sulfur emission restrictions have swapped the value of these two coals. Now low sulfur coal is more expensive than high sulfur coal.

When I worked for the Tennessee Valley Authority, I often traveled to the Paradise Steam Plant in Western Kentucky.[28] This plant was originally a mine-to-mouth operation. The coal field was literally on either side of the road leading into the plant. When environmental restrictions were tightened, the plant had to cut back on Kentucky (high sulfur) coal, in spite of having an elaborate processing system. To even things out, they burned Wyoming (low sulfur) coal part of the time.

The operators got a pitifully small bonus based on heat rate. The day shift guys had seniority over the night shift. The outcome was that the day shift got to run Kentucky coal and the night shift had to run Wyoming coal. On one occasion I had been working with a team for several days to perform a test of the cooling towers. We needed the weather to be within certain limits and also operation to be just right—most importantly full heat load. We had everything set up and ready just as the day shift ended and the night shift began. We had intended to work into the night and take advantage of the weather conditions.

Just as we began logging data, the heat load—indicated by the temperature drop across the cooling towers—fell significantly. We didn't know about the coal switch. We ran to the control room to fuss at the operator. He explained the situation and we were all very disappointed. They were feeding the low sulfur coal into the boilers as fast as the crushers, pulverizers, and conveyor belts could move it, but still couldn't make full load. In fact, they dropped 60MW on each of the two smaller units and 90MW on the large unit. That's four LM6000 gas turbines worth of power lost for burning low heating value fuel!

A limestone ($CaCO_3$) slurry in a wet scrubber is used to capture sulfur in the exhaust stream of some coal-fired boilers. The result is gypsum ($CaSO_4$), which has some industrial uses, but is often discarded.[29]

---

[28] There's even a famous country song about the place called, "Paradise," by singer-songwriter John Prine. The lyrics go... "Daddy, won't you take me back to Muhlenberg County; Down by the Green River where paradise lay? Well, I'm sorry my son, but you're too late in asking;; Cause Mister Peabody's coal train has hauled it away."

[29] Also from the tales of Paradise Steam Plant... They built a factory to make sheet rock and even gave the dry wall board away at one time until somebody noticed black specks and sued, forcing TVA to replace it all with new.

In our analysis of this reaction, we will consider a range of temperatures and pressures, various amounts of limestone, and various amounts of sulfur in the coal. The basic reaction is:

```
!6.3Air+Coal+0.2Cacarbonate=!Ar+H+Hdia+Hydroxl+Water
+Cdiox+Cmonox+N+Ndia+Ndiox+Nmonox+O+Odia+Sdiox+Smonox
+Cacarbonate+Cahydroxide+Casulfate+Casulfide+Casulfite
```

Coal and Air are defined in the database as:

```
     Air,[N0.7844 O0.2107 Ar0.0047 C0.0002],
        -34,45.77,6.33,9.76E-4,-9.96E-8,1,227,493
     Coal,[C0.5016 H0.4326 O0.0488 N0.0079 S0.0091],
               0,37.8,4.981,0,0,1,0,0
```

The basic results are:

```
Tp=2000K, Pp=1atm, Q/RTo=-405
substance  state     Y         X         Z    F    H/RTo   S/R    G/RTo
----------------------------------------------------------------------
Coal        !gas 1.0000000 0.1333333 1.00 1.00   14.3  25.81 -158.7
Air         !gas 6.3000000 0.8400000 1.00 1.00   22.5  30.46 -181.7
Cacarbonate !sol 0.2000000 0.0266667 0.01 0.10 -437.9  30.15 -640.0
----------------------------------------------------------------------
reactants        7.5000000 1.0000000 1.00 1.00   68.4 223.72 -1431.
======================================================================
Ar          !gas 0.0296100 0.0084480 1.00 1.00   14.3  28.14 -174.4
H           !gas 0.0000604 0.0000172 1.00 1.00  102.1  29.51  -95.7
Hdia        !gas 0.0004183 0.0001193 1.00 1.00   21.4  31.70 -191.1
Hydroxl     !gas 0.0030390 0.0008671 1.00 1.00   37.8  36.25 -205.2
Water       !gas 0.2143318 0.0611510 1.00 1.00  -68.1  34.67 -300.5
Cmonox      !gas 0.0061885 0.0017657 1.00 1.00  -21.3  37.67 -273.8
Cdiox       !gas 0.4966720 0.1417055 1.00 1.00 -122.7  38.86 -383.2
N           !gas 0.0000000 0.0000000 1.00 1.00  204.8  44.23  -91.7
Ndia        !gas 2.4706745 0.7049084 1.00 1.00   22.6  30.66 -183.0
Nmonox      !gas 0.0082674 0.0023588 1.00 1.00   59.6  38.90 -201.1
Ndiox       !gas 0.0000036 0.0000010 1.00 1.00   47.8  53.45 -310.6
O           !gas 0.0003441 0.0000982 1.00 1.00  115.5  33.61 -109.8
Odia        !gas 0.0662491 0.0189015 1.00 1.00   23.7  36.23 -219.3
Smonox      !gas 0.0000097 0.0000028 1.00 1.00   25.7  47.27 -291.2
Sdiox       !gas 0.0090900 0.0025935 1.00 1.00  -85.1  47.07 -400.6
Cacarbonate !sol 0.1999995 0.0570619 0.01 0.10 -437.9  30.08 -639.5
Cahydroxide !sol 0.0000002 0.0000001 0.01 0.10 -344.9  31.62 -556.8
Casulfide   !sol 0.0000000 0.0000000 0.01 0.10 -166.2  20.44 -303.3
Casulfite   !sol 0.0000000 0.0000000 0.01 0.10 -439.8  27.36 -623.2
Casulfate   !sol 0.0000003 0.0000001 0.01 0.10 -517.5  37.12 -766.4
----------------------------------------------------------------------
products         3.5049584 1.0000000 1.00 1.00 -105.6 112.86 -862.2
----------------------------------------------------------------------
difference      -3.9950416 0.0000000 0.00 0.00 -174.0 -110.9  569.2
Note: the ! above indicates non-ideal behavior
```

First, we vary temperature:

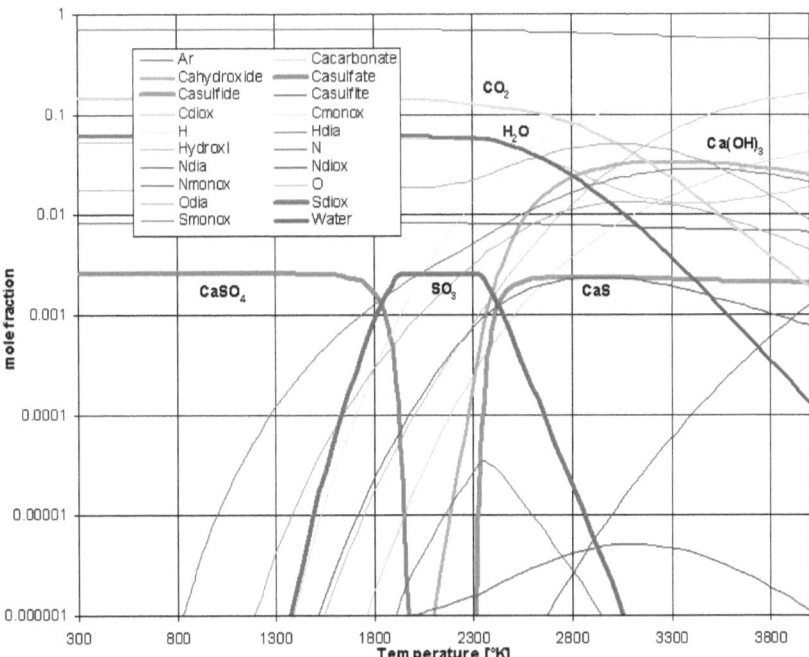

**Figure 42. Impact of Temperature on Sulfur Capture Using Limestone**

This is an excellent illustration of how conditions impact the outcome of chemical reactions. If we want to capture sulfur, we must keep the temperature below 1900°K or above 2400°K, as this is where the most $SO_2$ is produced (brown plateau). In order to produce gypsum, we must stay below 1900°K (the green curve falls off abruptly at this point). If we want to produce calcium sulfide (red CaS curve) or calcium hydroxide (orange $Ca(OH)_2$) curve, we must stay above 2400°K. For desulphurization of coal emissions, we must stay well below 1800°K. We also note that $H_2O$ (violet curve) falls off above 2400°K, as does $CO_2$ (sky blue), both of which are exothermic. For power (and steam), we would do best to keep the temperature below 2700°K.

Note that the Y-axis is logarithmic so that the fall off in $SO_2$ below 1900°K and above 2400°K is quite steep—orders of magnitude. As a wet slurry is nowhere near this temperature, the motivation for fluidized bed combustors becomes apparent. We can control the bed temperature to achieve the desired products. Of course, the optimum range will change with coal composition, air:fuel ratio, and lime:coal feed ratio, so this isn't a firm range for each and every design.

The OH:O ratio for this reaction is quite interesting...

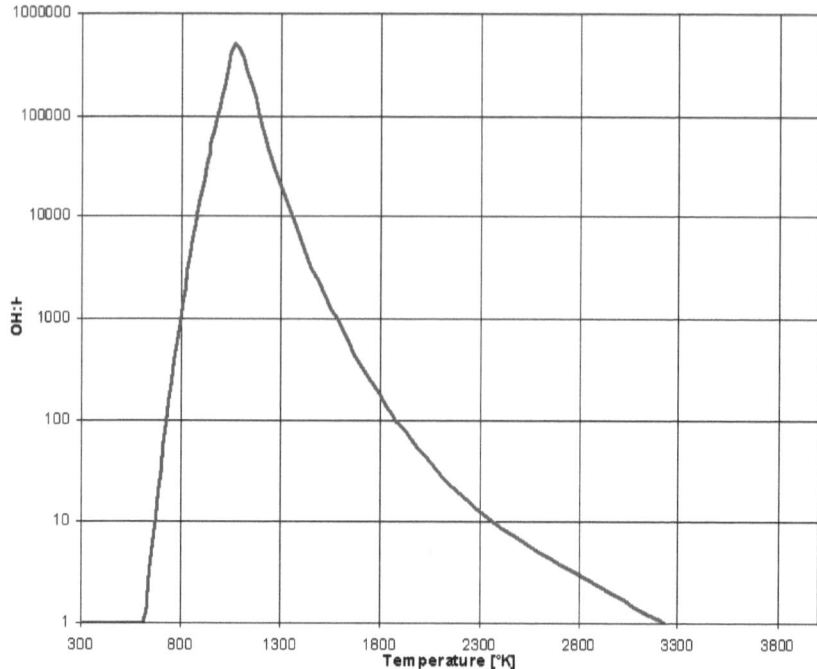

**Figure 43. Impact of Temperature on OH:H Ratio (Sulfur Capture)**

This is not surprising, considering the sharp increase in $Ca(OH)_2$. Controlling the pH for this reaction can prove a challenge, as can protecting equipment from attack. As this reaction clearly has three zones, we will consider the impact of pressure at each of these.

We see very little impact at 1400°K.

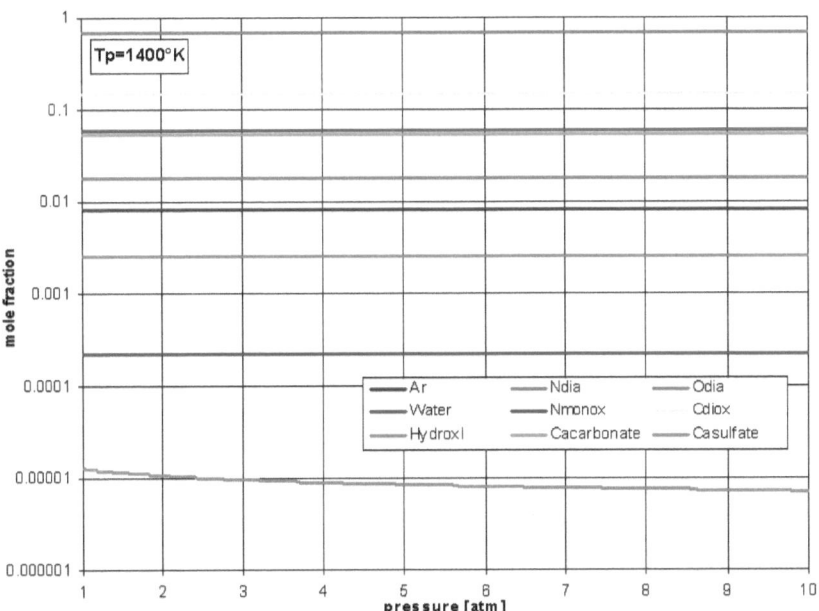

**Figure 44. Impact of Pressure on Sulfur Capture at 1400K**

The impact of pressure at 2150°K, in the center of the SO$_2$ plateau:

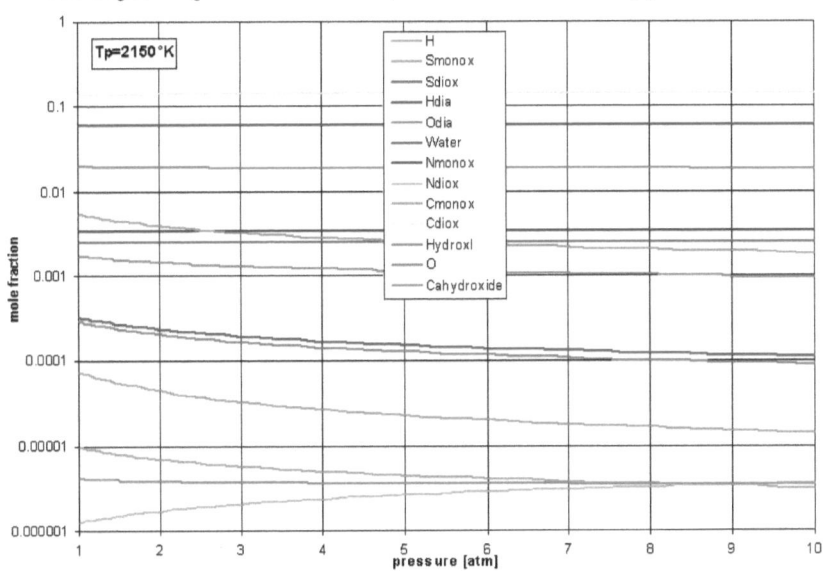

**Figure 45. Impact of Pressure on Sulfur Capture at 2150K**

The mix of important products (i.e., those more than 1:1,000,000) has changed from the previous plot, but there is little impact of pressure. At 2800°K:

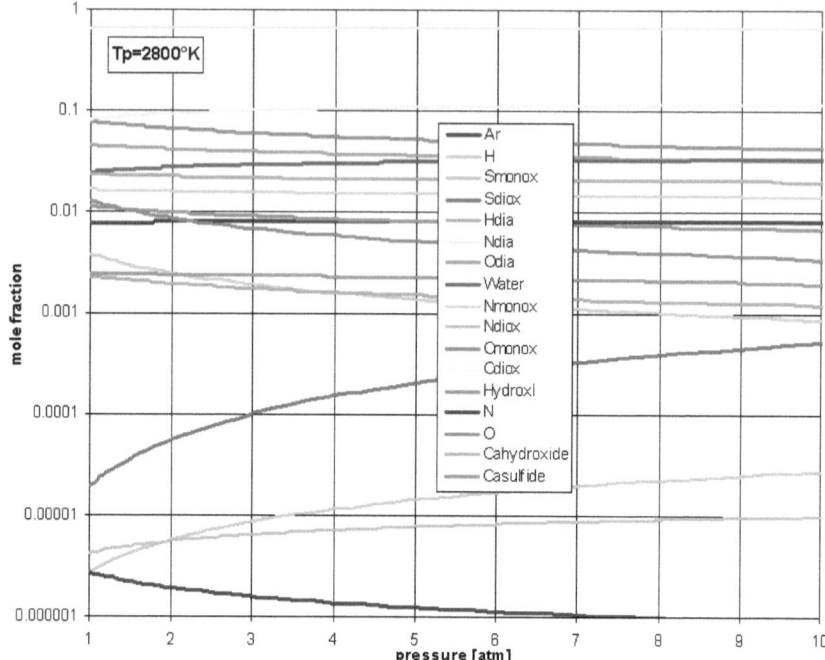

**Figure 46. Impact of Pressure on Sulfur Capture at 2800K**

There is some impact of pressure at 2800°K, as the products are responding more like gases. Still, this reaction is much more strongly dependent on temperature than pressure.

We will next consider air:fuel ratio.

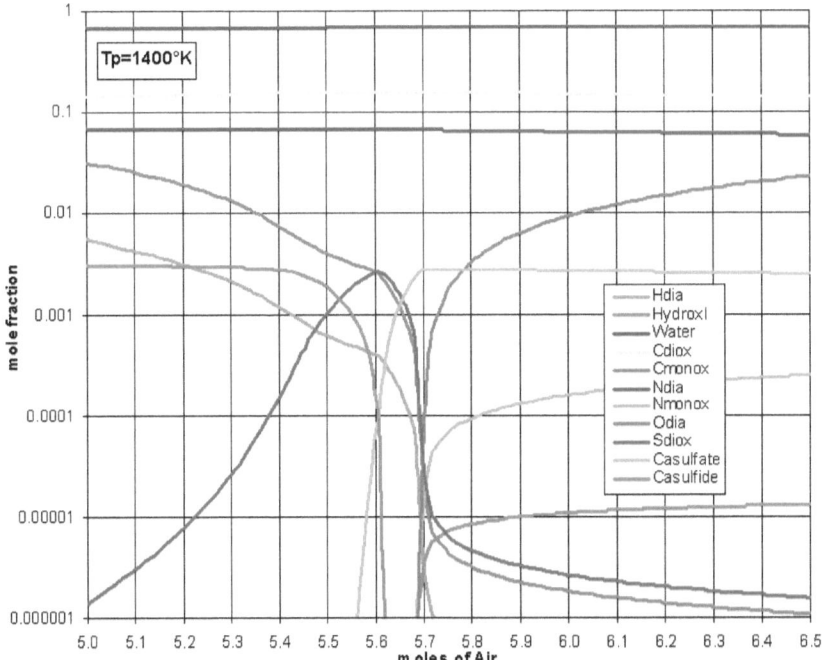

**Figure 47. Impact of Air/Fuel Ratio on Sulfur Capture at 1400K**

At 1400°K and 1 atm. we see a very strong transition at 5.7 moles of air per mole of coal. The peal $SO_2$ mole fraction occurs at 5.6 moles air. Gypsum rises abruptly and then levels off at 5.7 moles air. This is also the point where the pH changes most rapidly. CaS plummets at 5.6, never to rise again, not surprisingly with adequate oxygen. Unburned hydrocarbon also plummets at 5.7. Clearly, we would not want to operate below this air:fuel ratio.

At 2150°K and 1 atm. there is a transition at 5.4 moles of air, where the CaS falls off sharply and the $SO_2$ rises the SO peaks at 5.5 moles of air. $H_2O$ peaks at 4.5 moles of air and $CO_2$ peaks at 5.5 moles of air.

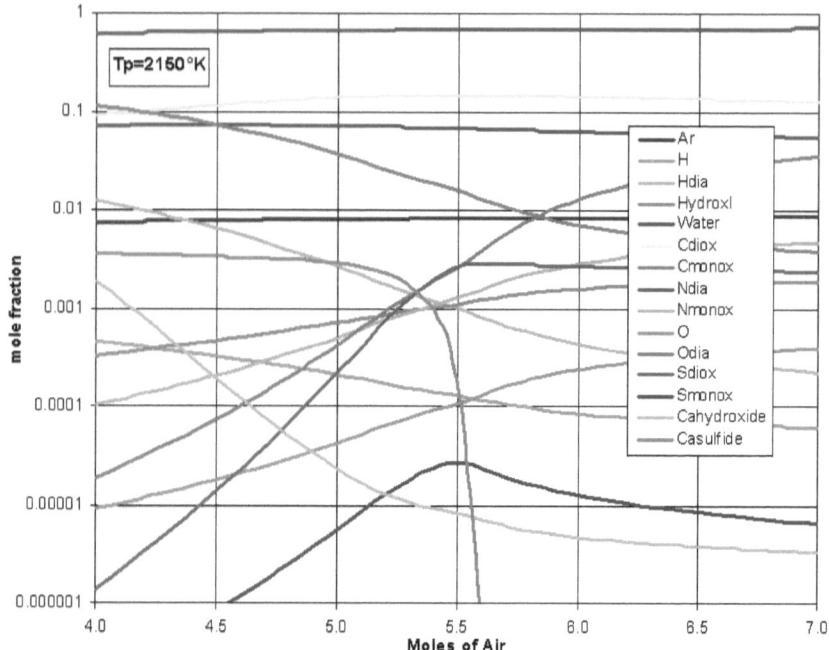

**Figure 48. Impact of Air/Fuel Ratio on Sulfur Capture at 2150K**

At 2800°K and 1 atm. there is no abrupt transition with air:fuel ratio:

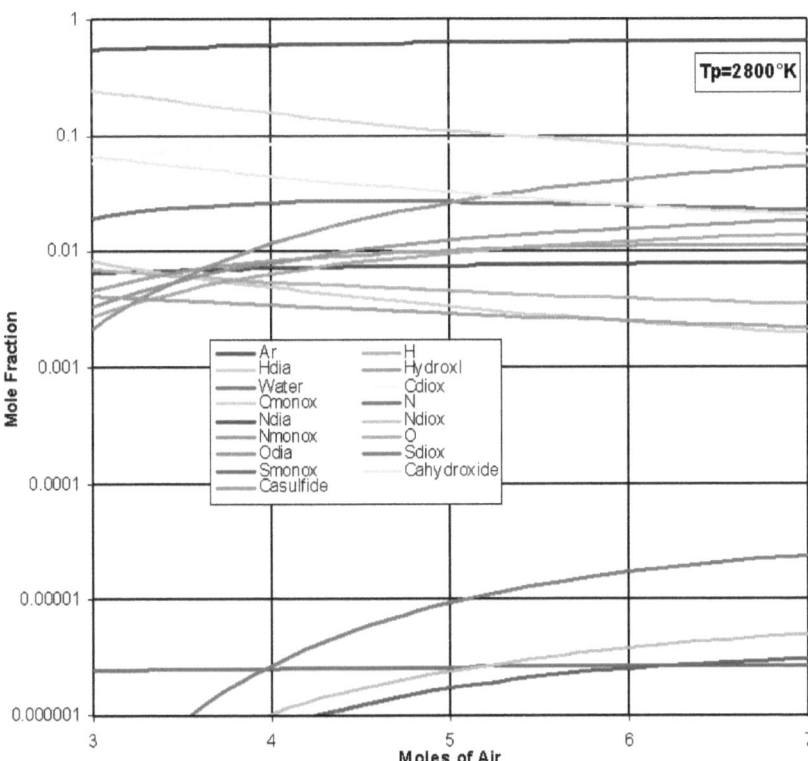

**Figure 49. Impact of Air/Fuel Ratio on Sulfur Capture at 2800K**

Confirmation

The last sequence of reaction plots, which combine gases and solids, demonstrates that the algorithms described in the previous chapters and implemented in the CREST computer program do not produce the same results, regardless of the conditions. Rather, the conditions and individual substance properties drive the calculations to diverse outcomes. While these particular numerical values are not compared to experimental results, they are consistent with anecdotal experience.

At 1400°K and 1 atm., even varying the CaCO3 moles by two orders of magnitude produces nothing remarkable. We conclude from this that, as long as you provide roughly enough, the exact feed rate isn't important.

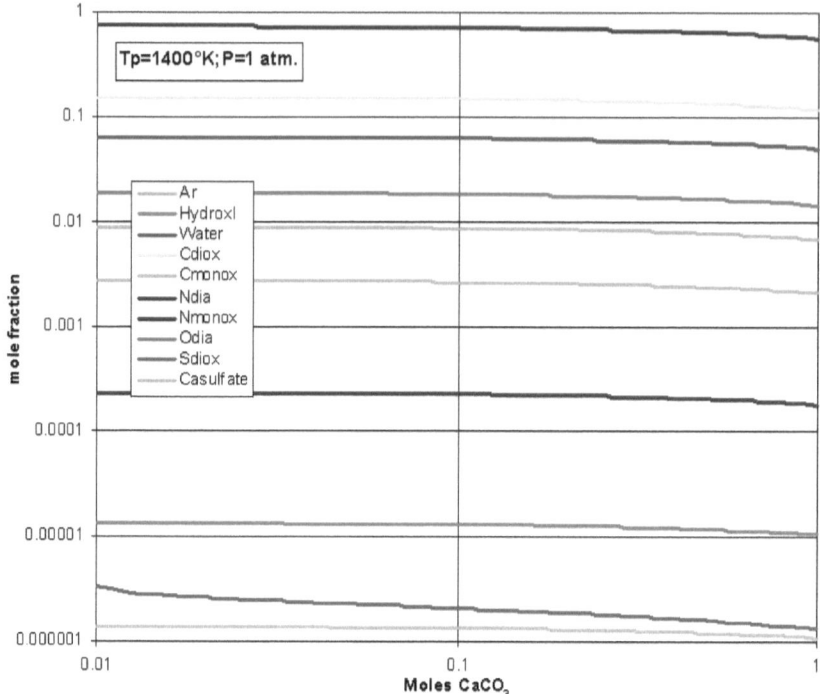

**Figure 50. Impact of CaCO$_3$ on Sulfur Capture at 1400K**

At 2150°K and 1 atm, there is still no transition revealed by changing the CaCO₃ moles by a factor of 100. Of course, the predominant products are different than at 1400°K and we are now in the SO₂ production range, but there is no critical molar value.

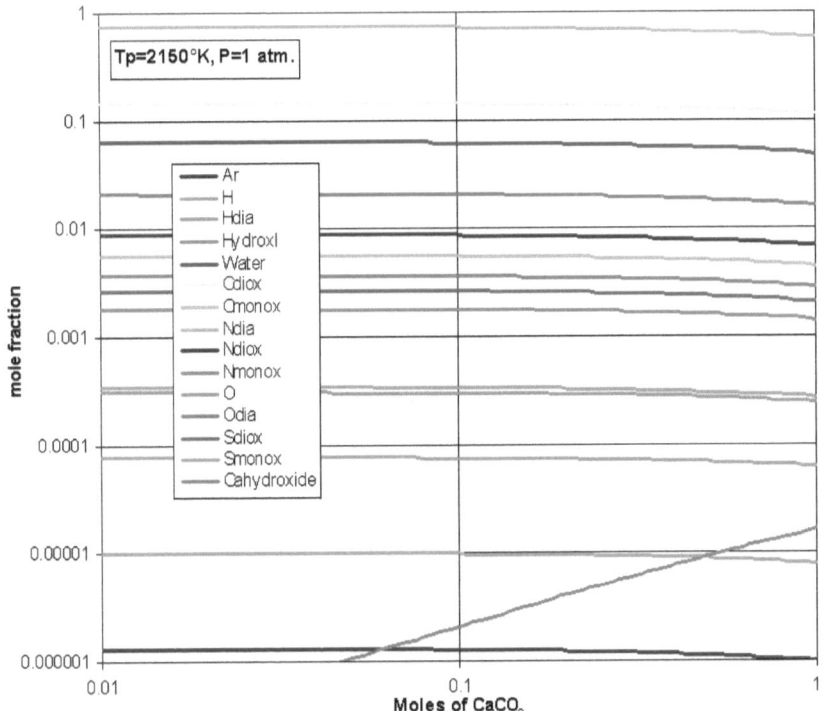

**Figure 51. Impact of CaCO3 on Sulfur Capture at 2150K**

At 2800°K and 1 atm., we see that $Ca(OH)_2$ production is calcium-limited and so is $SO_2$ and SO capture. Calcium hydroxide formation rises to a peak fraction of 0.07987 at 0.1265 moles $CaCO_3$ and then begins to drop, not due to further rising reactant abundance, but due to the total moles divisor. This is a gradual and proportional impact, not a critical point or abrupt transition.

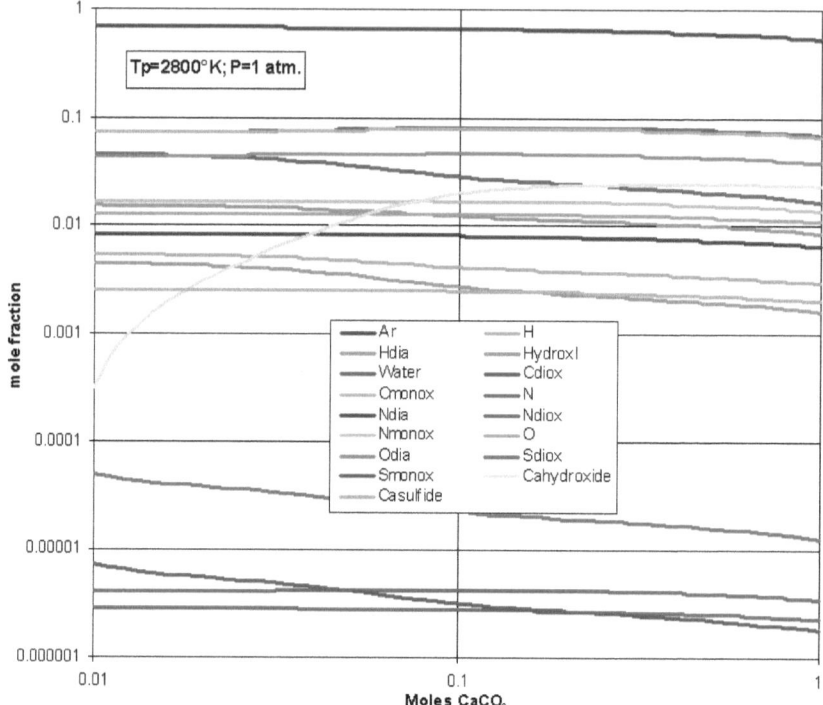

**Figure 52. Impact of CaCO3 on Sulfur Capture at 2800K**

## Chapter 13. Landfill Gas

One alternative energy source currently being utilized, which may potentially increase in the future is landfill gas. This is a much lower grade of fuel than conventional natural gas, which presents some challenges. For one thing, it has a considerably lower heating value, typically one-third to one-half and may be as low as one-fourth. Landfill gas also contains hydrogen sulfide ($H_2S$) that smells like rotten eggs and is toxic—so are the combustion products, including $SO_2$. On the positive side, it may be free—sort of, not entirely. You do have to capture, process, and utilize it—all of which costs money.

**Figure 53. Burning Off of Landfill Gas**

We will consider a baseline composition of: 50% methane, 40% carbon dioxide, 9% nitrogen, and 1% hydrogen sulfide. The basic reaction is:

```
!10Air+0.5 Methane+0.4 Cdiox+ 0.09 Ndia + 0.01 Hsulfide=
 !Ammonia+Ar+C+Cdiox+Cmonox+Cos+Cs+Ctriodia+Cyanide+H
   +Hdia+Hsulfide+Hydroxl+N+Ndia+Ndiox+Nitrousoxide
   +Nmonox+Nndiox+Nnmonox+Nnpentaox+Nntetraox+Nntriox
     +Ntriox+O+Odia+Ozone+Sdiox+Smonox+Striox
             +Sulfuricacid+Water
```

We can estimate the typical combustion temperature by starting with the reactants at 300°K (27°C) and 1 atm., setting the isobaric option, and Q/RT=0.

```
Tp=2048.2K, Pp=1atm, Q/RTo=0
substance     state      Y          X          H/RTo    S/R     G/RTo
-------------------------------------------------------------------------
Air          !gas  10.0000000  0.9090909    -0.0     23.15    -23.3
Methane      !gas   0.5000000  0.0454545   -30.2     33.79    -64.1
Cdiox        !gas   0.4000000  0.0363636  -158.6     29.03   -187.8
Ndia         !gas   0.0900000  0.0081818     0.0     27.86    -28.0
Hsulfide     !gas   0.0100000  0.0009091    -8.1     31.74    -40.0
-------------------------------------------------------------------------
reactants          11.0000000  1.0000000   -78.8    262.80   -343.0
=========================================================================
Ammonia      !gas   0.0000000  0.0000000     6.1     54.71   -369.5
Ar           !gas   0.0470000  0.0077869    14.7     28.28   -179.5
C            !gas   0.0000000  0.0000000   303.7     53.26    -62.0
Cdiox        !gas   0.8754305  0.1450395  -121.5     39.02   -389.4
Cmonox       !gas   0.0265695  0.0044020   -20.6     36.86   -273.6
Cos          !gas   0.0000000  0.0000000   -26.5     62.18   -453.4
Cs           !gas   0.0000000  0.0000000   115.5     61.65   -307.8
Ctriodia     !gas   0.0000000  0.0000000     9.5     78.19   -527.4
Cyanide      !gas   0.0000000  0.0000000    79.8     61.69   -343.8
H            !gas   0.0003702  0.0000613   102.5     28.30    -91.8
Hdia         !gas   0.0045627  0.0007559    22.1     29.95   -183.5
Hsulfide     !gas   0.0000000  0.0000000    16.0     55.25   -363.3
Hydroxl      !gas   0.0087140  0.0014437    38.5     35.84   -207.5
N            !gas   0.0000000  0.0000000   205.3     43.61    -94.2
Ndia         !gas   4.0070790  0.6638845    23.3     30.83   -188.4
Ndiox        !gas   0.0000024  0.0000004    48.9     54.56   -325.7
Nitrousoxide !gas   0.0000001  0.0000000    60.2     53.07   -304.1
Nmonox       !gas   0.0098390  0.0016301    60.3     39.37   -210.0
Nndiox       !gas   0.0000000  0.0000000   113.9     77.69   -419.9
Nnmonox      !gas   0.0000001  0.0000000    60.2     53.07   -304.1
Nnpentaox    !gas   0.0000000  0.0000000    72.1     93.37   -569.0
Nntetraox    !gas   0.0000000  0.0000000    58.2     83.93   -518.0
Nntriox      !gas   0.0000000  0.0000000    80.0     82.18   -484.2
Ntriox       !gas   0.0000000  0.0000000    61.7     70.69   -423.6
O            !gas   0.0005482  0.0000908   115.9     33.75   -115.8
Odia         !gas   0.0447967  0.0074218    24.4     37.28   -231.5
Ozone        !gas   0.0000000  0.0000000    85.2     62.99   -347.3
Sdiox        !gas   0.0099715  0.0016521   -84.3     47.64   -411.4
Smonox       !gas   0.0000269  0.0000045    26.3     46.89   -295.6
Striox       !gas   0.0000016  0.0000003  -130.6     57.76   -527.1
Sulfuricacid !gas   0.0000000  0.0000000  -241.7     82.89   -810.8
Water        !gas   1.0008952  0.1658262   -67.1     33.82   -299.3
-------------------------------------------------------------------------
products            6.0358077  1.0000000   -78.8    196.85  -1430.3
-------------------------------------------------------------------------
difference         -4.9641923  0.0000000    -0.0    -65.95  -1087.2
Note: the ! above indicates non-ideal behavior
```

The estimated temperature is 2048.2°K (1775°C)—pathetic, but it's cheap!

We first investigate the air:fuel ratio, holding the temperature and pressure constant. The range of mole fractions is much larger on this figure than on the preceding ones in order to show some of the nasties, including: HCN, COS, CO, SO, and $SO_2$. Groups of products are shown in similar colors, brown being reserved for cyanide.

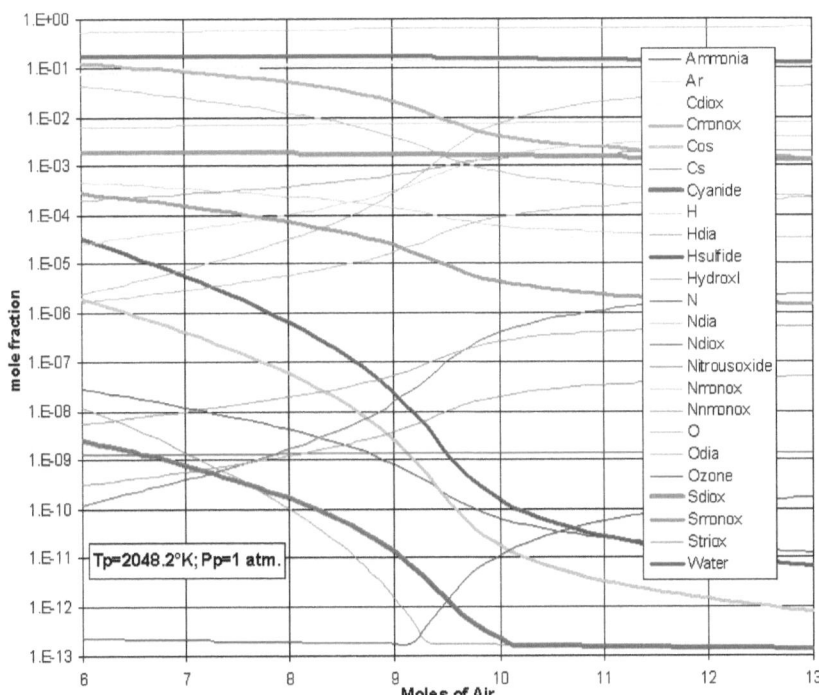

**Figure 54. Impact of Air/Fuel Ratio on the Combustion of Landfill Gas**

There is a clear transitional region at 9.5 moles of air, indicating the effective stoichiometric ratio for this fuel. $H_2O$ peaks at about 7 and $CO_2$ peaks at about 9.5 moles of air.

Holding pressure constant at 1 atm. and moles of air constant at 10, while varying temperature yields:

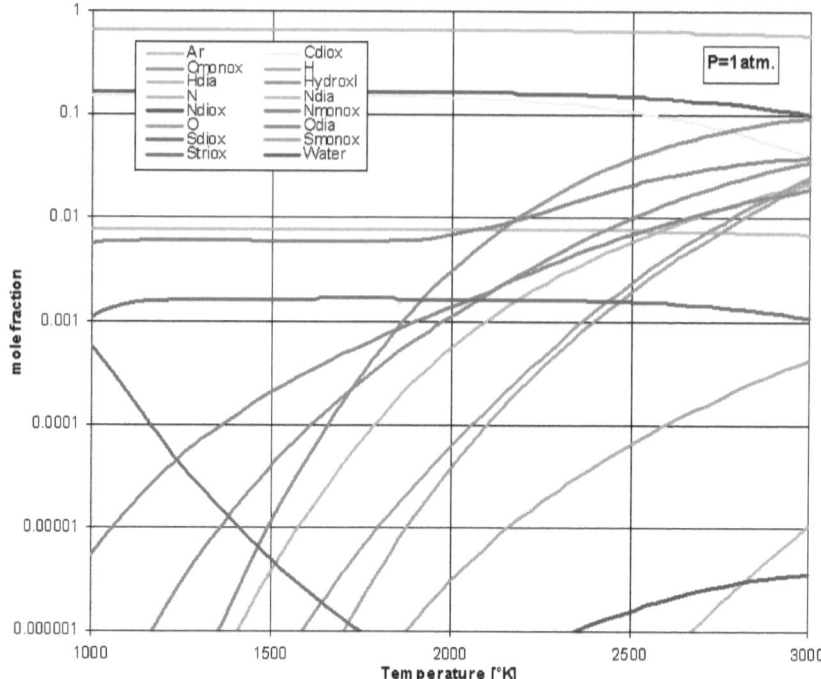

**Figure 55. Impact of Temperature on the Combustion of Landfill Gas**

$SO_2$ peaks at 1750°K, but SO and SO continue to rise with temperature. $H_2O$ and $CO_2$ also peak at about 1750°K, which would have the largest heat release. There is no critical temperature point or transitional temperature range for this reaction. Some provision must be made to control emissions if this fuel is used to any extent and it is typical for landfills.

The OH:H ratio for this fuel will also be a maintenance problem, as shown in this next figure:

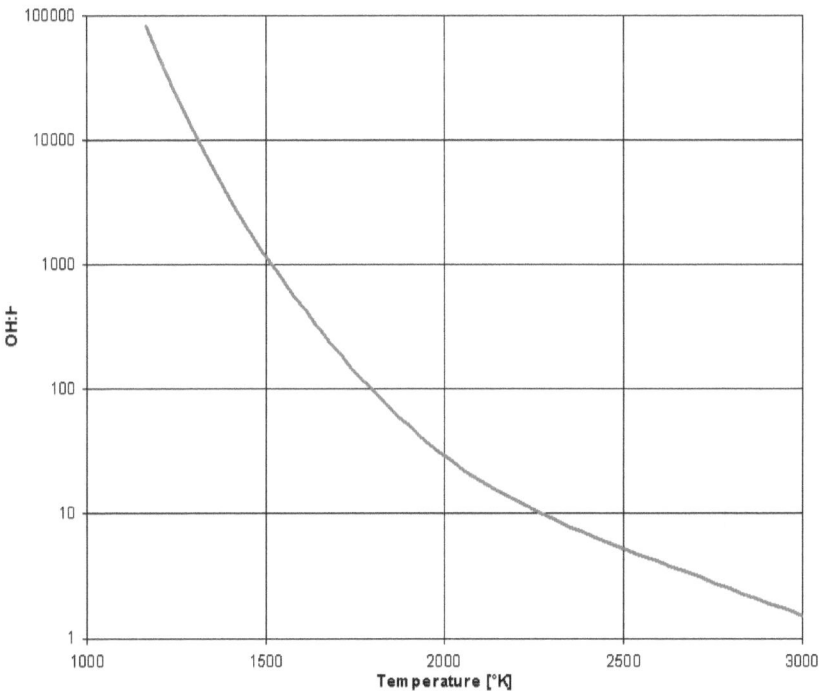

**Figure 56. Impact of Temperature on OH:H Ratio (Landfill Gas)**

There is a slight impact of pressure. CO and SO tend downward, which is to be expected with increased pressure favoring the larger compounds, $CO_2$, $SO_2$ and $SO_3$. There is no perceptible impact on $H_2O$ or $CO_2$.

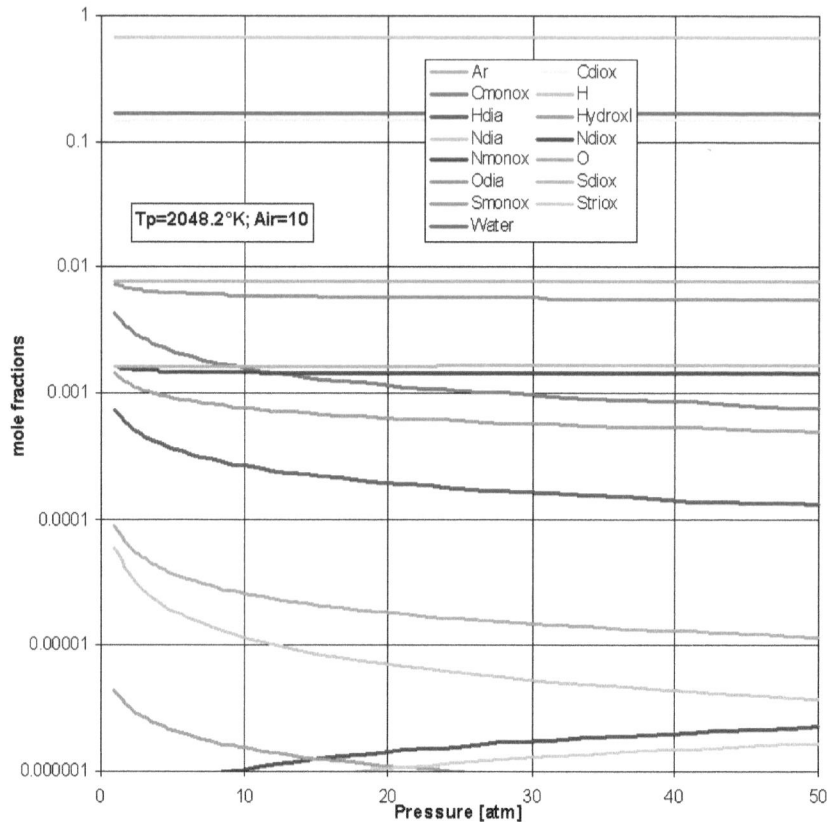

**Figure 57. Impact of Pressure on the Combustion of Landfill Gas**

# Appendix A. CREST Publication

I originally presented this material at the 1991 ASME Winter Annual Meeting in Atlanta, Georgia, when I was working for the Tennessee Valley Authority at the Engineering Laboratory in Norris. I have updated it for inclusion here, as it best describes development of the CREST program.

## Abstract

A method is presented by which chemical reactions can be conveniently expressed and automatically decomposed into the implied thermochemical equations. A variation of the RAND method with enhancements to the numerical algorithm that improve computational stability and convergence rate is used to solve these equations. A generalized equation of state capable of handling both vapor and liquid systems is used to obtain the partial fugacities and free energies. Results are presented for gaseous and aqueous systems.

## Nomenclature

A... square matrix containing $\partial g_I / \partial y_J$
B... column matrix containing $-g_I$
C... matrix containing elemental compositions
C... specific heat [J/gram-mole/K]
D... column matrix containing elemental abundances
g... specific Gibbs free energy [J/gram-mole]
G... Gibbs free energy of the system [J]
h... specific enthalpy of a component [J/gram-mole]
M... the number of components
N... the number of elements or irreducible members
s... specific entropy of a component [J/gram-mole/K]
y... component quantity [gram-moles]
Y... column matrix containing $y_J$
Greek
$\Phi$... quadratic function with linear constraints
$\varphi$... quadratic function
$\lambda$... Lagrange multipliers
$\Lambda$... column matrix containing $\lambda_I$
$\Omega$... column matrix containing linear constraints
Subscripts
0... reference or ground state
I... first index or row number
J... second index or column number

## Introduction

Chemical reactions are the controlling phenomena in many processes of interest to the utility and process industry. The primary method for analyzing

93

chemical reactions is to consider their ultimate or equilibrium state. The CREST (Chemical Reactions and Equilibrium Thermodynamics) computer code was developed to be a convenient and versatile means of determining chemical equilibria.

## The Gibbs Condition or Minimum Free Energy Postulate

The Gibbs condition or Minimum Free Energy Postulate is an extension of the Second Law of Thermodynamics and is equivalent to the condition of maximum entropy subject to the constraint of energy conservation or the First Law of Thermodynamics (van Wylen and Sonntag, 1973). Implicitly assumed in equilibrium analyses is the Erotic Surmise, which can be stated: the ultimate description of an equilibrium system is independent of time (Pierce, 1968). The Gibbs condition can then be summarized: the tendency is for reactions to proceed such that the free energy is reduced, the final condition being that of minimum free energy.

Throughout this development it will be assumed that the thermodynamic systems in question are "open" (viz. maintained at constant pressure). In order to apply the developments to a system that is "closed" (viz. maintained at constant volume) all references to the Gibbs free energy and enthalpy can be replaced by Helmholtz free energy and internal energy, respectively. This is done automatically by the CREST computer program.

## Solution of Chemical Equilibria

The equilibrium chemical reaction problem is defined as satisfying the Gibbs condition subject to the element abundance[30] and non-negativity[31] constraints. The Gibbs condition is nonlinear; whereas the element abundance constraints are linear. The strategy employed here is a variation of the RAND method (White, et al., 1958), where the unknowns are the molar abundances. The resulting nonlinear system is solved in an iterative manner as a sequence of linearized systems which satisfy the non-negativity constraint at every iteration and the element abundance constraint at most iterations (excepting those which would violate the non-negativity constraint), leaving the free energy extrema to be approached only as the iterations converge.

## Formulation

The formulation begins with the mathematical expression for the Gibbs condition and the elemental abundance constraints. As with the RAND

---

[30] Abundance of elements, reactants, or products is the quantity (or the number of moles) present in the system. This is different from concentration, which might be moles per unit volume, and mole fraction, which is the abundance of one species divided by the total number of moles.

[31] The non-negativity constraint is the same as recognizing that one cannot have negative moles of something.

algorithm, these constraints are imposed through the use of Lagrange multipliers. The resulting set of nonlinear simultaneous equations is solved using a hybrid method, which will be described subsequently.

## The Gibbs Condition

The Gibbs free energy of a system is the sum of the product of the molar abundances and the specific free energies of the components, or:

$$G = \sum y_I g_I \tag{A.1}$$

The extrema (and in particular the minima) of G are located at the point(s) where the gradient of G is zero (i.e., where the partial derivatives of G with respect to the molar abundances, $y_I$, are zero). This can be expressed as:

$$\frac{\partial g_I}{\partial y_J} = 0 \tag{A.2}$$

Substituting the definition of G, or Equation A.1, into Equation A.2 yields a set of nonlinear simultaneous equations, which can be expressed in matrix form:

$$AY - B = 0 \tag{A.3}$$

where Y is the column matrix containing the molar abundances, $y_I$, A is the square matrix having the elements

$$A_{IJ} = \frac{\partial g_I}{\partial y_J} \tag{A.4}$$

and B is the column matrix containing the free energy of the components, or:

$$B_I = -g_I \tag{A.5}$$

Equation A.3 is nonlinear because the elements of the matrix (i.e., the component specific free energies and their partial derivatives) depend on several factors, including the molar abundances, $y_I$.

## The Element Abundance and Non-Negativity Constraints

The constraints conserving the abundance of the N elements (or irreducible members—which could be ions) of the system can be expressed as a system of N simultaneous linear equations involving the M molar abundances, or:

$$CY - D = \Omega \rightarrow 0 \tag{A.6}$$

where C is a rectangular matrix having N rows and M columns and D and $\Omega$ are column matrices having M elements. The non-negativity constraints can be expressed by:

$$y_I > 0 \tag{A.7}$$

## Application of Lagrange Multipliers

The method of Lagrange multipliers is the typical way of solving constrained extrema (Wylie, 1975). In order to implement this method, it is necessary to define a function whose solution is the desired extrema (i.e., the Gibbs condition). The quadratic function whose solution is Equation A.3 can be expressed in matrix form by Equation A.8 (Ortega and Rheinboldt, 1970).

$$\phi = Y^T A Y - Y^T B \qquad (A.8)$$

Defining the Lagrange multipliers,

$$\Lambda = \left[ \lambda_I \right]^T \qquad (A.9)$$

the element abundance constraints can be added to form the constrained function:

$$\Phi = \phi + \Lambda^T \Omega = Y^T A Y - Y^T B + \Lambda^T \left[ CY - D \right] \qquad (A.10)$$

The extrema of the function $\Phi$ occur when the partial derivatives with respect to the molar abundances, $y_I$, and the Lagrange multipliers, $\lambda_I$, are all zero. This condition can be expressed by the following partitioned matrix, having N+M rows and columns.

$$
\left[ \begin{array}{c|c} A & C^T \\ \hline C & 0 \end{array} \right]
\; x \;
\left[ \begin{array}{c} Y \\ \hline \Lambda \end{array} \right]
\; = \;
\left[ \begin{array}{c} B \\ \hline D \end{array} \right]
$$

## Method of Solution

Equation A.11 is the essence of what Smith and Missen (1982) refer to as the "RAND algorithm." White, Johnson, and Dantzig (who were working for the RAND Corporation at the time) introduced the same basic equations in 1958, although they came to them through a different approach. They began with an expression for the free energy of a mixture of ideal gases, applied in their words the "Method of Steepest Descent", and used the first term in a Taylor's expansion of the mixture free energy to arrive at essentially the same equations. In the terminology of Smith and Missen, what White, Johnson, and Dantzig developed was a second order, non-stoichiometric, algorithm using the Newton-Raphson method of solution. Formulating these equations is one matter; developing a practical manner in which to solve them is quite another. Smith and Missen (1982) and Cruise (1964) provide some helpful flow charts which illustrate the basic steps involved.

## Solving the Simultaneous Equations

In Equation A.11 matrices A and B are nonlinear, while C and D are linear. The original RAND method is basically a straight Newton iteration. Newton's method is not always stable or rapid. An adaptive hybrid of the Newton, Steepest Decent, and Conjugate Gradient methods is used to make the solution more robust. This method is described by Benton (1991); however, the necessity and basis for it can well be gleaned from Ortega and Rheinboldt (1970), Powell (1977), More and Sorensen (1984), and Fletcher (1987).

## Computation of the Free Energy

How one evaluates the elements of matrix A determines whether the method is restricted to a mixture of ideal gases—as was the case with White, et al.—or whether it can be applied to real systems composed of real substances. The mathematics are the same until actually evaluating Equation A.4 (i.e., the partial derivatives of the free energy with respect to the molar abundances). The effort, which must be invested in evaluating these partials, depends on the accuracy required and the degree of nonideality exhibited by the system. The CREST computer code was developed in such a way that it provides varying levels of complexity. Some substances are treated as ideal gases while others are not. The mixing rule can also be selected.[32]

## Equation of State and Mixing Rule

Equations of state and mixing rules abound in the literature. The reasons for selecting one over the others vary almost as much as the equations and rules themselves. The current selection was based on the criteria that there be exactly three roots of any subcritical isotherm, that Maxwell's Criterion[33] hold, that all the critical properties match exactly, that an excessive number of empirical data are not required, and that the equations be reasonably accurate not only for "well-behaved" substances but for a group of substances chosen for their notorious "ill" behavior: water, ammonia, carbon dioxide, and neon.

The cubic equation of state described by Fuller (1976) was found to be the most satisfying in regards to these criteria. Maxwell's Criterion is used to infer the saturation pressure from the equation of state.[34] As stated previously, opinions vary considerably as to the selection of equations of state. Abbott (1973) provides a very enlightening discussion of the advantages and limitations of cubic equations of state. The correspondence by Chung, Hamam, and Lu (1977) in which they discuss the greater accuracy of the method which Chung and Lu developed—provided one is not so concerned about polar molecules— illustrates these varying preferences and perspectives.

---

[32] A mixing rule is a model for the behavior of a mixture--specifically how the properties of the individual species are influenced by the presence of the other species in the mixture.

[33] Satisfying Maxwell's Criterion is equivalent to requiring continuous saturation phase fugacity.

[34] Another way of stating Maxwell's Criterion is that the area under a subcritical isotherm (on a pressure-volume diagram) is equal to the saturation pressure times the difference in phase volumes.

Mixing rules also vary considerably. Smith and Missen (1982), Prausnitz (1969), and Stadler (1989) provide detailed discussions of mixing rules. The mixing rule currently used by the CREST code is that of Redlich and Kwong (1949) as it is easily implemented and provides a significant level of improvement over the ideal mixture model. Eventually the method described by Liu, Wimby, and Gren (1989) will be implemented as an option.

### Initial Estimates of the Molar Abundances

There is also the necessity of obtaining initial estimates of the molar abundances, $y_I$, and satisfying the non-negativity constraints. The SIMPLEX method (cf. Wagner, 1975) is used to accomplish this. The molar abundances are bounded below by a small positive number (for practical purposes this needs to be somewhere between the square and cube root of the smallest number greater than zero that the computer can store) and bounded above by the abundance of the most restrictive irreducible constituent (e.g., the upper bound on $CO_2$ would be the smaller of the total number of moles of C and half that of O). Matrices A and B can be linearized in order to obtain the initial estimates by computing secant partials at the lower and upper bound.

### Implementation

A computer code, CREST, was developed in order to implement these algorithms. The substances participating in the reaction and their properties as well as how their state is to be treated and their mixture computed is defined for CREST in a symbolic form, which is interpreted based on syntax.

### The Use of Syntax to Define Substances and Reactions

Syntax is used not only to identify the species but also to distinguish between irreducible, conserved participants (i.e., elements or ions) and reducible participants (i.e., compounds). The elemental (and electron) abundance constraints can be determined automatically from the composition of the compounds, which is specified, in a user-defined database. Table 1 is an abbreviated excerpt from a CREST database. In the interest of brevity, not all of the information for each substance is listed in the table. This other information includes such things as the temperature dependence of the constant pressure specific heat at zero pressure, $Cp0$, the standard state, and critical properties if known. The thermodynamic and transport properties in the absence of any chemical reactions can be accessed through a separate computer code, FEAST (Benton, 1987).

The reaction to be solved by defined by entering it in symbolic form interactively or in a file as illustrated in Equation A.12. The number of moles of the reactants must be specified (or assumed to be 1). Of course, the number of moles of products must not be specified, as these are the solution sought after. Various conditions such as temperature, pressure, and heat transfer and constraints such as adiabatic, isobaric, and isometric can also be prescribed interactively, in a batch file, or progmatically. The equation of state to be used for each species as well as the mixing rule is also specified by the syntax. The equation of state is specified by a special character preceding the name of the substance and the mixing rule is specified by a special character preceding the equals sign in the reaction.

$$\text{Acetylene}+2.7\text{Odia}=\text{Water}+\text{Cmonox}+\text{Cdiox} \\ +\text{H}+\text{Hdia}+\text{O}+\text{Odia}+\text{Hydroxl}+...$$

(A.12)

## Programmatic Control

The method is implemented in the form of a computer code, which can be run interactively, from a batch file, or spawned as a subtask from within another process with complete programmatic control. A source code written in C is supplied with the CREST executable, which illustrates how the program can be spawned within a loop or loops in order to model sequential reactions, variation of reactant species, and produce custom graphical output.

## Program Performance

Solution of 72 simultaneous reacting species in an aqueous solution requires between 4 and 12 minutes on a PC (20 MHz 80386/7)[35] depending on the mixing model used. Solution of 100 simultaneous reacting species in a gaseous mixture requires between 10 and 30 minutes on such a machine depending on the mixing model used. Simple reactions (e.g., 20 species in an ideal gas mixture) require only a few seconds. Most of the CREST code was written in FORTRAN,[36] except for the I/O, full-screen editor, reaction interpreter, and matrix solver, which were written in assembler.

## Program Results

Figure A.1 shows the computed variation with temperature of 20 product species resulting from the combustion of coal. Figure A.2 shows the computed variation with oxygen of 14 product species resulting from the combustion of a hydrocarbon in the presence of sulfur. Figure A.3 shows the computed variation with hypochlorite of 23 product species resulting from the corrosion of cement containing 3 variants of asbestos in an aqueous solution. Every step from the reaction to the hard copy of these plots was provided by the program CREST. Certainly many more examples could be given. More complex reactions (e.g., sequential steps such as occur in the presence of a catalyst and more species) have been analyzed using the program; but this increased complexity is difficult to present graphically and concisely.

Table 1. Excerpts from a CREST Database

```
*define elements (order is immaterial)
*name h0 s0 Cp0 ... other ... weight
```

---

[35] HA! Those were the days... now it takes only seconds.
[36] CREST has been translated into C and is an interactive Windows® application.

```
C 0 1.4 6.05 ................ 12.011 carbon
H 93800 27.4 4.97 ........... 1.008 monatomic
    hydrogen
O 107000 38.5 5.24 .......... 15.999 monatomic
    oxygen
*define compounds (order is immaterial)
*name [composition] h0 s0 Cp0 ...other...
Hdia [H2] 0 31.2 6.53 .................
Odia [O2] 0 49.0 6.69 .................
Acetone [CH3COCH3] 93200 70.5 17.90 ...
Acetylene [C2H2] 97500 48.0 10.50 ......
Mek [CH3COC2H5] 101000 80.8 24.60 .....
Cmonox [CO] 47600 47.3 6.71 ...........
Cdiox [CO2] 169000 51.1 8.01 ..........
Water [H2O] 104000 45.1 6.87 ..........
```

Model Comparison

The CREST computer code is similar in concept to the STANJAN computer code, which was developed by Prof. William Reynolds of Stanford University. The basic methodology of finding the minimum free energy subject to the elemental abundance and non-negativity constraints is the same. Both codes are interactive, make use of an expandable database of substances, and run on IBM compatible PCs. CREST, however, has several features not present in STANJAN. These include: optional real gas equations of state, nonideal mixing rules, a built-in full-screen editor, graphical output to 12 different devices, batch or complete programmatic control, and all of the matrix operations are coded in assembler for maximum speed. The addition of non-ideal mixing rules alone represents a considerable increase in complexity.

## Closing

The rather complicated task of determining thermochemical equilibria has been implemented in the form of a computer code, which is relatively simple to use. The code (CREST) has been available on the ASME/CIME computer in one form or another since 1988 and has been satisfactorily used to model a variety of problems including: flue gas desulphurization, ammonia injection into flue gas, calcium oxide production, and calcium leaching from wetted fiber reinforced cement products.

## References

Abbott, M. M., 1973, "Cubic Equations of State," AIChE Journal (19:596 601).

Benton, D. J., 1987, "FEAST: Fast Estimation of Thermodynamic and Transport Properties," a computer code available on the ASME/CIME Bulletinboard (608)233 3378.

Benton, D. J., 1988, "CREST: Chemical Reactions and Equilibrium Statistical Thermodynamics," loc. cit.

Benton, D. J., 1991, "Applications of a Hybrid Derivative-Free Algorithm for Locating Extrema," SIAM-SEAS, Cullowhee, NC.

Chung, W. K., S. E. M. Hamam, and B. C.-Y. Lu, 1977, Letter to the Editor regarding paper by Fuller (1976), op. cit. 16:494 495.

Cruise, D. R., 1964, "Notes on the Rapid Computation of Chemical Equilibria," Journal of Physical Chemistry (68:3797 3802).

Fletcher, R., 1987, *Practical Methods of Optimization*, John Wiley and Sons, New York, NY.

Fuller, G. G., 1976, "A Modified Redlich-Kwong-Soave Equation of State Capable of Representing the Liquid State," Ind. Eng. Chem. Fund., 15:254 257.

Liu, Y., M. Wimby, and U. Gren, 1989, "An Activity-Coefficient Model for Electrolyte Systems," Computers in Chemical Engineering (13:405 410).

More, J. J. and D. C. Sorensen, 1984, "Newton's Method," Studies in Numerical Analysis, G. H. Golub, ed., The Mathematical Association of America, pp. 29 82.

Ortega, J. M. and W. C. Rheinboldt, 1970, *Iterative Solution of Nonlinear Equations in Several Variables*, Academic Press, New York.

Pierce, F. J., 1968, *Microscopic Thermodynamics*, International Textbook Company, Scranton, Pennsylvania.

Powell, M. J. D., 1977, "Restart Procedures for the Conjugate Gradient Method," Mathematical Programming, Vol. 12, pp. 241 254.

Prausnitz, J. M., 1969, *Molecular Thermodynamics of Fluid-Phase Equilibrium*, Prentice-Hall, Englewood Cliffs, New Jersey.

Redlich, O. and J. N. S. Kwong, 1949, "On the Thermodynamics of Solutions," Chemical Review (44:233 244).

Smith, W. R. and R. W. Missen, 1982, *Chemical Reaction Equilibrium Analysis: Theory and Algorithms*, John Wiley and Sons, New York.

Stadler, H. P., 1989, *Chemical Thermodynamics*, The Royal Society of Chemistry, Cambridge, U.K. (available in the U.S. through CRC Press).

van Wylen, G. J. and R. E. Sonntag, 1973, *Fundamentals of Classical Thermodynamics*, John Wiley and Sons, New York.

Wagner, H. M., 1975, *Principles of Operations Research, 2nd Ed.*, Prentice-Hall, Englewood Cliffs, New Jersey.

White, W. B., S. M. Johnson, and G. B. Dantzig, 1958, "Chemical Equilibrium in Complex Mixtures," Journal of Chemical Physics (28:751 755).

Wylie, C. R., 1975, *Advanced Engineering Mathematics*, 4th Ed., McGraw-Hill, New York.

## Appendix B. CREST Computer Program

Before describing the CREST computer program, I must tell you the story of how it came to be. The snowstorm of February 1986 stranded me and hundreds of other travelers inside Chicago's O'Hare Airport for three days. We were unfortunate enough to be in the process of transferring from one of the big airlines to a small commuter line. This left us between carriers with no one obligated to make any provision for our comfort—no hotel room, bed, or food—just a few vending machines, which were soon empty. These were restocked once or twice, but quickly ravaged until the travelers' cash or stocks were depleted. All this took place at one of the many satellite "gates" or waiting areas in the vast maze, which is O'Hare. Fortunately, I had one of the very early portable computers: a Compaq "lunchbox" model. With nothing better to do, I set out to solve a problem that had been festering for years: chemical reactions.

### Coding

CREST was originally written in FORTRAN. If C existed at that time, it wasn't available on my machine running DOS™. I didn't translate the code into C until September 1991. Parsing the reactions in FORTRAN was dreadful. The C language is far superior to anything else and so much better than FORTRAN for handling strings and characters. I didn't migrate CREST from a console application (often called a DOS box) to Windows® until February 2007, because there was no pressing need. I didn't add SI units until March 2010.

CREST is now entirely written in C, will compile for 32-bit or 64-bit operation without modification, and uses no libraries or DLLs besides those that are part of the Windows® operating system. CREST will run on any 32-bit or 64-bit version of Windows®, including: 95, 98, ME, 2K, XP, Vista, 8, 9, and 10. CREST does not require any service pack or any version of Internet Explorer®. CREST does not use the vile .Net™ Framework in any way, shape, or form.

### Formulation

CREST implements the formulations presented in Chapters 1 through 4. The Gibbs free energy is minimized for isobaric reactions and the Helmholtz free energy is minimized for polybaric reactions. The Method of Steepest Descent is used. The element abundances are also conserved using Lagrange Multipliers so that the entire process is nonlinear constrained minimization.

All reactants and products default to ideal gas behavior unless real fluid behavior is activated, as described subsequently. When activated, the free energy and its derivatives are corrected using the formulas presented in Chapters 5 through 7. The Redlich-Kwong equation of state us used along with the associated mixing rules and weighted critical properties, as described.

Because the Hessian (i.e., the matrix A containing the second partial derivatives of the total free energy with respect to the molar abundances) becomes ill-conditioned when the compressibility, Z, or fugacity coefficient, φ, falls below unity, a rank-one adjustment is made ($y_I = y_I$) to maximize the condition number and achieve convergence. This modification to the method is effective, even for very small values of Z and φ.

## Operation

CREST is an interactive Windows® application and is launched like any other. The program is button-driven, as illustrated below:

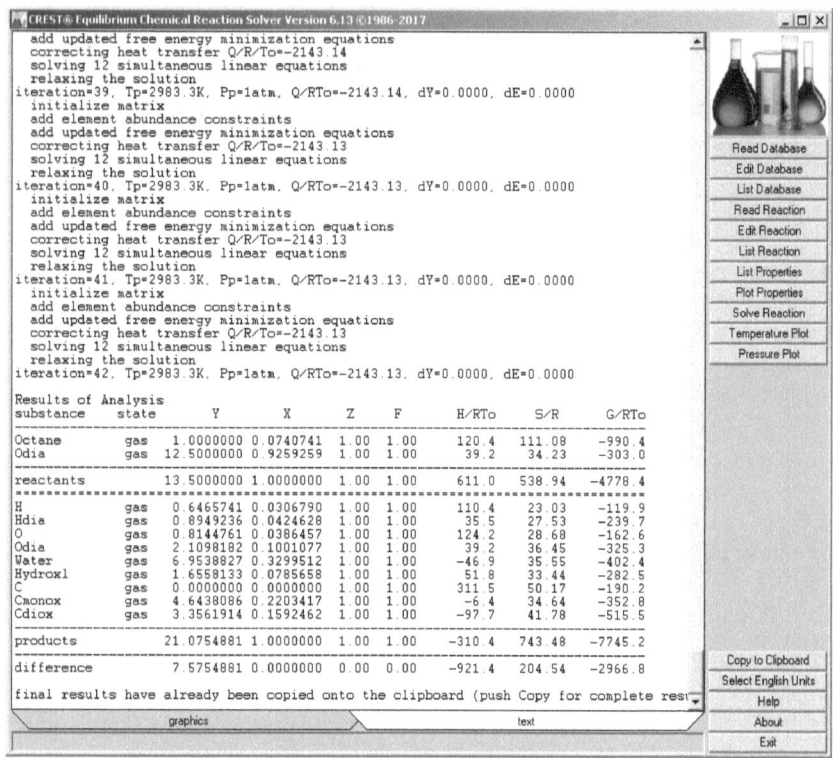

**Figure 58. CREST Screenshot Showing Reaction Details**

The first step is to load one of the two databases: gases or aqueous. You can build your own databases. Databases are stored in CSV format. Databases can be edited with Excel® or notepad™.

The second step is to load a reaction. Several are provided and you can create your own. Reactions are stored in .TXT format. Reactions can be edited with notepad™. DO NOT use MSWord® or anything like it to edit a text file!

After that, you can list, solve, or plot the reaction. The tabular results are automatically copied onto the clipboard and can be pasted directly into Excel®.

You cannot paste data into CREST. The program has two tabs: text and graphics. If you press the Copy to Clipboard button, this will copy the currently selected tab onto the clipboard.

You can create four different types of plots. One example is:

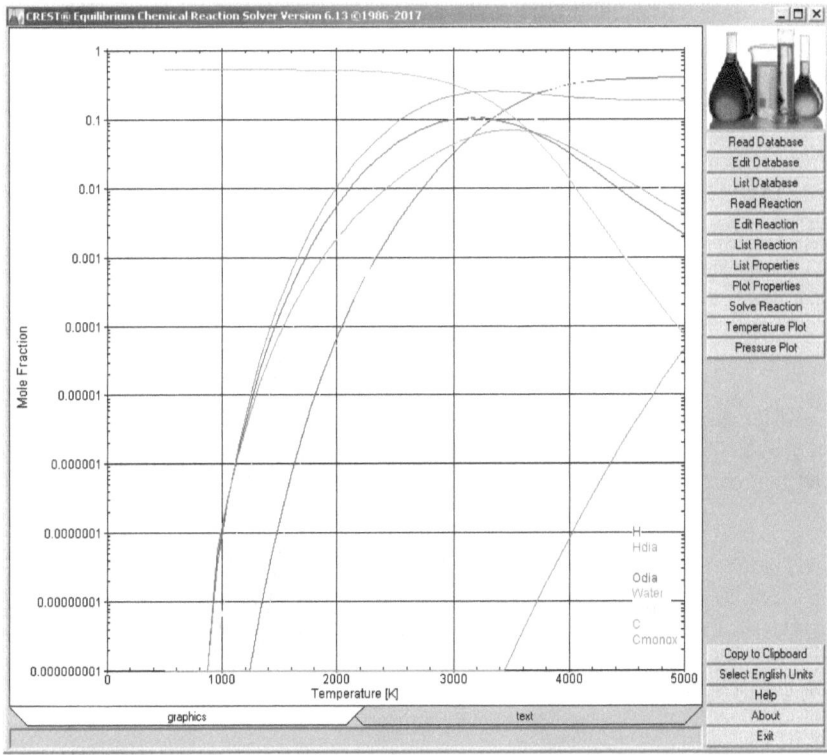

**Figure 59. CREST Screenshot Showing Graphical Results**

### Elements and Compounds

Elements have a name, molecular weight, specific heat (1 to 3 coefficients), a heat of formation, and an entropy of formation. Compounds have a name, a composition, specific heats, a heat of formation, and an entropy of formation. Compounds must be composed entirely of elements, which must have already been defined. Compounds do not have a molecular weight, as this is calculated from the composition and corresponding elements. Ions can either be elements (if composed of a single atom) or compounds (if composed of more than one atom).

Element and compound names must be unique. These can be the same as in the periodic table, but need not be. Names must begin with an upper case letter [A-Z] and may continue on with one or more lower case letters [a-z]. Names may also contain a trailing carat (^) or tilde (~), but may not begin with these. Names must not contain numbers (0-9) or other symbols, including plus (+) or minus (-). For instance, diatomic oxygen might be Odia, but CANNOT be O2. The most common liquid on Earth could be Water, but CANNOT be H2O. Names may not contain parentheses. The purpose of these restrictions is disambiguation. Text files do not have subscripts or superscripts, so there's no way to tell the difference between H2O, $H_2O$, and $H^2O$. I am also not going to interpret parentheses, such as $Mg(OH)_2$ or $Al_2(SO_4)_3$.

Compound compositions are enclosed in square braces, such as:

Water [H2O]

Hydrogenperoxide [H2O2]

Acetone [CH3COCH3]

Elements can appear more than once in a compound composition statement, such as is the case with acetone above. The number of atoms always follows the element name and need not be an integer, as in the following examples:

Air [N0.7844 O0.2107 Ar0.0047 C0.0002]

Coal [C0.5016 H0.4326 O0.0488 N0.0079 S0.0091]

Spaces are not required. Do not use exponential notation, as "1.23E+34" means something other than $1.23 \times 10^{34}$ and will be interpreted as 1.23 moles of E +34 moles of something else.

## Reactions

Reactions begin with one or more elements or compounds, optionally preceded by numbers. These are the reactants, which are followed by equals (=) and the products. Products can be elements or compound and may be separated by plus (+), but this is not necessary. Product names must not be preceded or followed by any numbers, as this is why you're solving the reaction to determine. There must not be any numbers on the right side of the equals (=).

## Ideal vs. Real Properties

All elements and compounds are presumed to exhibit ideal gas behavior. You can optionally activate real (i.e., non-ideal) properties for any reactant or product by following the name with an exclamation point (!). You can use this control of each item to investigate what difference it might make in the outcome. If you precede all reactants or all products with an exclamation point (!), this will activate real properties for everything on that side of the equals (=). If critical pressure and temperature are not provided in the database, these will be estimated. In either case, critical properties are necessary in order to implement the Redlich-Kwong mixing rules, which apply whenever real properties are activated.

## Validation

The following combustion graph of a hydrocarbon fuel having an average composition of $C_{10}H_{22}O_{40}$ at 1 atm. from the literature[37] is provided as evidence that the CREST program works. These calculations assume ideal gases, so the real gas corrections have not been activated for this example.

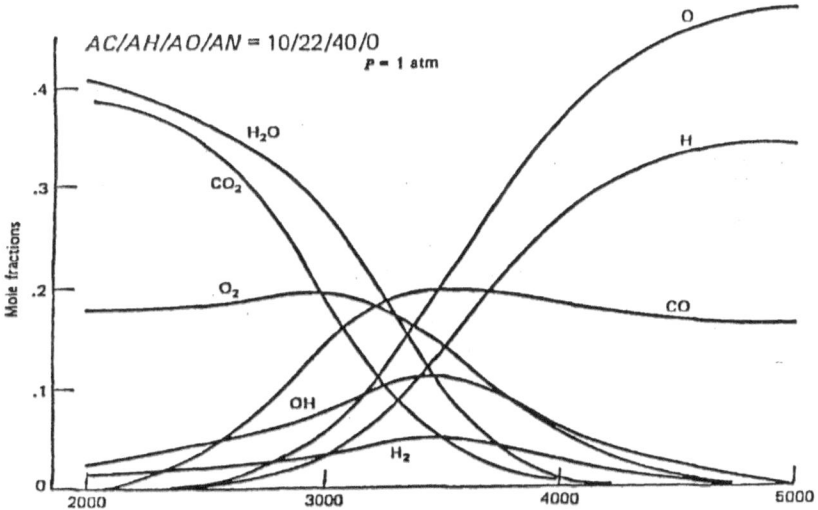

**Figure 60. Published Mole Fractions vs. Temperature (prior)**

[37] Again, I must apologize for not recalling where this figure came from. Like the two previous ones, this has been in a loose-leaf 3-ring binder for the past three decades.

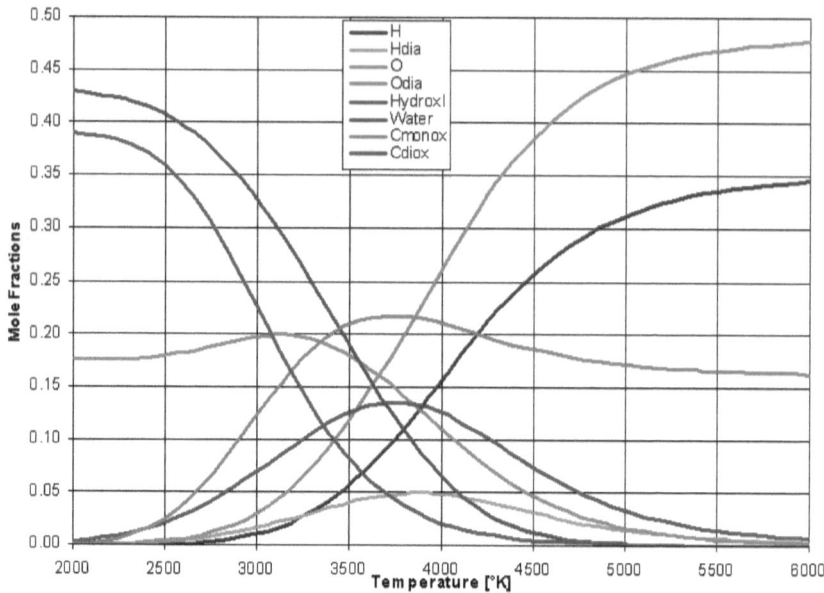

**Figure 61. Calculated Mole Fractions vs. Temperature (this work)**

## also by D. James Benton

*3D Articulation: Using OpenGL*, ISBN-9798596362480, Amazon, 2021 (book 3 in the 3D series).

*3D Models in Motion Using OpenGL*, ISBN-9798652987701, Amazon, 2020 (book 2 in the 3D series.

*3D Rendering in Windows: How to display three-dimensional objects in Windows with and without OpenGL*, ISBN-9781520339610, Amazon, 2016 (book 1 in the 3D series).

*A Synergy of Short Stories: The whole may be greater than the sum of the parts*, ISBN-9781520340319, Amazon, 2016.

*Azeotropes: Behavior and Application*, ISBN-9798609748997, Amazon, 2020.

*bat-Elohim: Book 3 in the Little Star Trilogy*, ISBN-9781686148682, Amazon, 2019.

*Boilers: Performance and Testing*, ISBN: 9798789062517, Amazon 2021.

*Combined 3D Rendering Series: 3D Rendering in Windows®, 3D Models in Motion, and 3D Articulation*, ISBN-9798484417032, Amazon, 2021.

*Complex Variables: Practical Applications*, ISBN-9781794250437, Amazon, 2019.

*Compression & Encryption: Algorithms & Software*, ISBN-9781081008826, Amazon, 2019.

*Computational Fluid Dynamics: an Overview of Methods*, ISBN-9781672393775, Amazon, 2019.

*Computer Simulation of Power Systems: Programming Strategies and Practical Examples*, ISBN-9781696218184, Amazon, 2019.

*Contaminant Transport: A Numerical Approach*, ISBN-9798461733216, Amazon, 2021.

*CPUnleashed! Tapping Processor Speed*, ISBN-9798421420361, Amazon, 2022.

*Curve-Fitting: The Science and Art of Approximation*, ISBN-9781520339542, Amazon, 2016.

*Death by Tie: It was the best of ties. It was the worst of ties. It's what got him killed.*, ISBN-9798398745931, Amazon, 2023.

*Differential Equations: Numerical Methods for Solving*, ISBN-9781983004162, Amazon, 2018.

*Equations of State: A Graphical Comparison*, ISBN-9798843139520, Amazon, 2022.

*Evaporative Cooling: The Science of Beating the Heat*, ISBN-9781520913346, Amazon, 2017.

*Forecasting: Extrapolation and Projection*, ISBN-9798394019494, Amazon 2023.

*Heat Engines: Thermodynamics, Cycles, & Performance Curves*, ISBN-9798486886836, Amazon, 2021.

*Heat Exchangers: Performance Prediction & Evaluation*, ISBN-9781973589327, Amazon, 2017.

*Heat Recovery Steam Generators: Thermal Design and Testing*, ISBN-9781691029365, Amazon, 2019.

*Heat Transfer: Heat Exchangers, Heat Recovery Steam Generators, & Cooling Towers*, ISBN-9798487417831, Amazon, 2021.

*Heat Transfer Examples: Practical Problems Solved*, ISBN-9798390610763, Amazon, 2023.

*The Kick-Start Murders: Visualize revenge*, ISBN-9798759083375, Amazon, 2021.

*Jamie2: Innocence is easily lost and cannot be restored*, ISBN-9781520339375, Amazon, 2016-18.

*Kyle Cooper Mysteries: Kick Start, Monte Carlo, and Waterfront Murders*, ISBN-9798829365943, Amazon, 2022.

*The Last Seraph: Sequel to Little Star*, ISBN-9781726802253, Amazon, 2018.

*Little Star: God doesn't do things the way we expect Him to. He's better than that!* ISBN-9781520338903, Amazon, 2015-17.

*Living Math: Seeing mathematics in every day life (and appreciating it more too)*, ISBN-9781520336992, Amazon, 2016.

*Lost Cause: If only history could be changed...*, ISBN-9781521173770, Amazon, 2017.

*Mass Transfer: Diffusion & Convection*, ISBN-9798702403106, Amazon, 2021.

*Mill Town Destiny: The Hand of Providence brought them together to rescue the mill, the town, and each other*, ISBN-9781520864679, Amazon, 2017.

*Monte Carlo Murders: Who Killed Who and Why*, ISBN-9798829341848, Amazon, 2022.

*Monte Carlo Simulation: The Art of Random Process Characterization*, ISBN-9781980577874, Amazon, 2018.

*Nonlinear Equations: Numerical Methods for Solving*, ISBN-9781717767318, Amazon, 2018.

*Numerical Calculus: Differentiation and Integration*, ISBN-9781980680901, Amazon, 2018.

*Numerical Methods: Nonlinear Equations, Numerical Calculus, & Differential Equations*, ISBN-9798486246845, Amazon, 2021.

*Orthogonal Functions: The Many Uses of*, ISBN-9781719876162, Amazon, 2018.

*Overwhelming Evidence: A Pilgrimage*, ISBN-9798515642211, Amazon, 2021.

*Particle Tracking: Computational Strategies and Diverse Examples*, ISBN-9781692512651, Amazon, 2019.

*Plumes: Delineation & Transport*, ISBN-9781702292771, Amazon, 2019.

*Power Plant Performance Curves: for Testing and Dispatch*, ISBN-9798640192698, Amazon, 2020.

*Practical Linear Algebra: Principles & Software*, ISBN-9798860910584, Amazon, 2023.

*Props, Fans, & Pumps: Design & Performance*, ISBN-9798645391195, Amazon, 2020.

*Remediation: Contaminant Transport, Particle Tracking, & Plumes*, ISBN-9798485651190, Amazon, 2021.

*ROFL: Rolling on the Floor Laughing*, ISBN-9781973300007, Amazon, 2017.

*Seminole Rain: You don't choose destiny. It chooses you*, ISBN-9798668502196, Amazon, 2020.

*Septillionth: 1 in $10^{24}$*, ISBN-9798410762472, Amazon, 2022.

*Software Development: Targeted Applications*, ISBN-9798850653989, Amazon, 2023.

*Software Recipes: Proven Tools*, ISBN-9798815229556, Amazon, 2022.

*Steam 2020: to 150 GPa and 6000 K*, ISBN-9798634643830, Amazon, 2020.

*Thermodynamic and Transport Properties of Fluids*, ISBN-9781092120845, Amazon, 2019.

*Thermodynamic Cycles: Effective Modeling Strategies for Software Development*, ISBN-9781070934372, Amazon, 2019.

*Thermodynamics - Theory & Practice: The science of energy and power*, ISBN-9781520339795, Amazon, 2016.

*Version-Independent Programming: Code Development Guidelines for the Windows® Operating System*, ISBN-9781520339146, Amazon, 2016.

*The Waterfront Murders: As you sow, so shall you reap*, ISBN-9798611314500, Amazon, 2020.

*Weather Data: Where To Get It and How To Process It*, ISBN-9798868037894, Amazon, 2023.